THE

HISTORICAL RECORD

OF THE

27TH INNISKILLING REGIMENT,

FROM THE PERIOD OF ITS INSTITUTION AS A VOLUNTEER CORPS TILL THE PRESENT TIME;

WITH AN APPENDIX,

ROLL OF COLONELS, ROLL OF PRESENT OFFICERS, ETC., ETC.

BY W. COPELAND TRIMBLE.

TO

THE RIGHT HONOURABLE

WILLIAM WILLOUGHBY,

THIRD EARL OF ENNISKILLEN,

COLONEL OF THE FERMANAGH LIGHT INFANTRY MILITIA,

NEPHEW OF GENERAL SIR GALBRAITH LOWRY COLE, G.C.B., ETC., ETC., ETC.

(THE MOST DISTINGUISHED COLONEL OF THE 27TH INNISKILLING

REGIMENT OF FOOT, COMMANDER OF THE FOURTH

DIVISION DURING THE PENINSULAR WAR);

AND DESCENDANT OF SIR WILLIAM COLE, KNT., THE FOUNDER OF THE TOWN

This Volume,

COMPILED DURING LEISURE HOURS,

TREATING OF THE SERVICES OF THE INNISKILLING REGIMENT

FROM THE PERIOD OF ITS BEING FIRST RAISED

AS A VOLUNTEER CORPS, TILL THE PRESENT TIME,

IS RESPECTFULLY INSCRIBED.

PREFACE.

The Historical Records of the 27th or Inniskilling Regiment of Foot were first published by the present writer in a series of nineteen chapters in the *Fermanagh Reporter*, in the latter end of 1874 and beginning of 1875, whilst the head-quarters of that distinguished regiment lay in Enniskillen. At the request of some of the commissioned and non-commissioned officers, and of others interested in the Inniskillings, I have been induced to republish the regiment's history, with various and considerable additions procured at the Horse Guards, for which I am indebted to the kindness of the Adjutant-General Sir Richard Airey, and to Viscount Crichton, M.P. The publication of the work has been entrusted to the firm of Messrs. W. Clowes and

Sons, in order that the present volume may form one of the series issued by them for Mr. Cannon, late of her Majesty's War Office.

In the compilation of this History of the 27th Regiment, I have had the indispensable aid of the regimental records (to which I was permitted access by Colonel Freer), the perusal of the documents at the Horse Guards, and the advantage of information on the subject gathered by Mr. Cannon, who had contemplated the publication of the work which it now falls to the lot of the humble subscriber to discharge. The additional assistance of local knowledge, and an acquaintance with the regiment by historical and present associations, have contributed to its favour in no slight degree.

To those gentlemen of the Press who kindly noticed with commendatory words—and passed over the weaknesses of—the first publication of the Records, and to those of the public who were so good as to approve of my efforts, I beg to tender my grateful acknowledgments. Youth and inexperience cannot hide from me the obligations I owe them, when in the present day so many candidates for popular approval, much worthier than I, fail to receive the same.

Nothing tends more to maintain the *esprit de corps* of the army than a knowledge of past renown. In this respect few regiments, if any, can boast of more glorious deeds or honourable service for sovereign and country, than the foot regiment of my native town. The tale of their achievements for nearly 200 years creditably sustains the fame of the British soldier, and succeeding sovereigns have transmitted their acknowledgments to posterity by inscribing a bead-roll of glory on their victorious banners.

Enniskillen has reason to be proud of her favourite corps, for as long as the national spirit survives, so long will be remembered the brilliant achievements and intrepid valour of the gallant 27th. In the most glorious page of the eventful history of England's many wars will be found the distinguished services of an Irish regiment whose bravery never flinched—the splendid record of lofty self-devotion, sterling pluck, and unconquerable courage.

Whatever its failings, in one respect the work is unique: it details the history of the regiment to the present time. For it I claim no literary merit, but hope that the "plain, unvarnished tale" will suffice to gratify the reader's taste. In a letter to

Mr. Cannon, dated July 25th, 1852, from the Royal Barracks, Dublin, Lieutenant-Colonel Cunynghame correctly observed: "The real truth is, that the history of the regiment is in a great degree the history of England since the day of William III."

W. COPELAND TRIMBLE.

ENNISKILLEN, *October*, 1875.

CONTENTS.

APPENDIX.

LIST OF ILLUSTRATIONS.

"Shoulder Arms"

UNIFORM 27TH REGIMENT ENNISKILLEN 1875.

W.F. WAKEMAN. DEL.

For Trimble's Record.

PENI
OULOUS
WATE
ST LUCIA
NIVELLE ORTHES
PYRENEES
EGY
NEC ASPER
HONOURS
27TH
INNISK

NSULA
MAIDA
RLOO
BADAJOZ
SALAMANCA
VITTORIA
PT.
A TERRENT.
OF THE
ILLING
7
REGT.

UNIFORM OF 27TH REGIMENT INNISKILLINGS 1742.

(OBTAINED AT HORSE GUARDS)

For Trimble's Record of the 27th.

ADDITIONAL NAMES, RECEIVED TOO LATE FOR ALPHABETICAL ARRANGEMENT.

Alexander, Joseph, Esq., Enniskillen.
Baxter, H. A., Esq., London, Ontario, Canada.
Bradshaw, John A., Esq., Professor of History, College of Madras, and Fellow of Madras University.
Cockburn, P. C., Esq., Shangana Castle, Bray.
L'Estrange, G. C., Esq., Carlton Club, London, Sheriff of Fermanagh.
Russell, J. Creagh, Esq., Clarinda Park, Kingstown.

LIST OF SUBSCRIBERS.

Archdale, Wm., Esq., M.P., D.L., Riversdale, Co. Fermanagh.
Archdale, Captain Hugh M., late 52nd Light Infantry.
Arnold, Matthew, Esq., Professor of Music, Enniskillen.
Andrews, J., Private, 27th Inniskillings.
Allen, W. ditto ditto

Belmore, Right Hon. the Earl of, Castlecoole, Enniskillen.
Blake, P. J., Esq., Q.C., Chairman of Fermanagh.
Baumgartner, Major-General Robert J., late 27th Inniskillings.
Brien, John Dawson, Esq., J.P., Castletown, Monea.
Brownrigg, Captain R. W., 27th Inniskillings.
Barton, Edward, Esq. ditto
Bayly, J. C., Esq. ditto
Bray, N. A., Esq. ditto
Bond, A. G. M., Esq. ditto
Brennan, Quartermaster C. ditto
Bennett, J. A., Esq. ditto
Bradshaw, Rev. W. H., Kilskeery Rectory, Co. Tyrone.
Barnes, E., Sergeant, 27th Inniskillings.
Bourke, P. ditto ditto
Byrne, T. ditto ditto
Brunson, W., Corporal ditto
Byrne, W. ditto ditto
Blake, T., Private ditto
Bowes, J. ditto ditto
Brown, P. ditto ditto
Brunning, J. ditto ditto
Barkway, W. ditto ditto
Brennan, J. ditto ditto
Bowman, J. ditto ditto
Byrne, T. ditto ditto

CRICHTON,* Viscount, M.P., Crom, Newtownbutler.
COLE, Viscount, Cheshire.
CUNYNGHAME, Lieutenant-General, late 27th Inniskillings.
COLE, Colonel Hon. H. A., M.P., late 6th Inniskillings.
COLE, Colonel A. Lowry, C.B., Wimbledon.
COWELL, Major H., 27th Inniskillings.
CAINE, Captain H. M. ditto
COFFEY, Captain F. ditto
COTTON, Captain J. M. V., late of ditto
COLLUM, Captain William, late 94th Regiment.
CARSON, Wm., Esq., T.C., Willoughby Place, Enniskillen.
COLHOUN, James, Esq., *Sentinel*, Londonderry.
CASSIDY, A., Esq., T.C., Enniskillen.
CALVERT, O., Colour-Sergeant, 27th Inniskillings.
COBBE, C., Sergeant ditto
CRAPNELL, R., Corporal ditto
CUE, W. H., ditto ditto
CONOLLY, T., Private ditto
CARSON, J. J. ditto ditto
CORDUCK, G. ditto ditto
CLINTON, J. ditto ditto
CONOLLY, P. ditto ditto
CLARKE, T. ditto ditto
CARRICK, M. ditto ditto
CUSHING, E. ditto ditto
CARROLL, T. ditto ditto
CHATTON, G. ditto ditto
CUTTING, H. G. ditto ditto
CAHILL, M. ditto ditto
CHAMBERS, J. ditto ditto
CLARKE, T. ditto ditto

DANE, Paul,† Esq., Enniskillen and Ballyshannon.
DUFFIN, J. Corporal, 27th Inniskillings.
DOWLING, J. ditto ditto
DRUM, T., Private ditto
DUNCAN, W. ditto ditto
DRENNAN, J. ditto ditto
DICKSON, J. ditto ditto

ENNISKILLEN, Right Hon. Earl of, Florence Court.
ERNE, Right Hon. Earl of, K.P., Crom Castle.
ELLIS, W., Private, 27th Inniskillings.

* Descendant of Captain Crichton, 1688-89.
† Descendant of Paul Dane, Provost 1689.

FREER, Colonel Richard, commanding 27th Inniskillings.
Fermanagh Club, per Dr. Walsh, J.P.
FRAMINGHAM, E., Colour-Sergeant, 27th Inniskillings.
FARMER, S., Private ditto

GEDDES, Major A. D., 27th Inniskillings.
GRANT, A. H., Esq. ditto
GREER, Rev. Samuel, A.B., Rector of Enniskillen.
GUNNING, John, Esq., Enniskillen.
GORDON, Wm. Edward, Esq. Enniskillen.
GRIMES, W., Colour-Sergeant, East York Militia, late 27th Regiment.
GORDON, T., Private, 27th Inniskillings.
GLADWELL, W. ditto ditto
GRIBBON, T. ditto ditto
GRIMES, J. ditto ditto
GOODRICH, T. ditto ditto
GRAHAM, G. ditto ditto

HALL, Mrs., 12, Kensington Gate, late Inismore Hall.
HERRING, Captain W. 27th Inniskillings.
HALES, Captain A. ditto
HARE, C. W., Esq. ditto
HANDYSIDE, W. H., Colour-Sergeant ditto
HOWLETT, F. W., Sergeant ditto
HARDY, H. A., Private ditto
HOLLAND, T. ditto ditto
HARDY, A. ditto ditto
HENDERSON, T. ditto ditto

IRVINE,* Colonel J. G., D.L., commanding Fermanagh Light Infantry, Killadeas, Co. Fermanagh.
IRVINE, Edward, Esq., J.P., Derrygore, Enniskillen.
IRVINE, A. J., Esq., 27th Inniskillings.

JENKINS, J. D., Sergeant, 27th Inniskillings.
JOHNSTON, W. ditto ditto
JONES, J., Private ditto
JOHNSTONE, C. ditto ditto

KERR, Captain T. M., 27th Inniskillings.
KETTLEWELL, W. W., Esq., late 27th Inniskillings.
KELLY, P., Paymaster-Sergeant, Armagh Militia, late 27th Regiment.
KERR, J., Pioneer-Sergeant, 27th Inniskillings.
KANE, M., Private ditto
KEELAN, M. ditto ditto

* Descendant of Sir Gerard Irvine, 1688.

LANESBOROUGH, Right Hon. Earl of, Lanesborough Hall.
Library, 6th Inniskillings.
LOWRY, J., Colour-Sergeant, 27th Inniskillings.
LEGGETT, W., Corporal ditto
LYNESS, J., Private ditto
LEWIS, H., Private, 27th Inniskillings.
LEATHAM, R. ditto ditto
LENNON, J. ditto ditto
LISLE, W. ditto ditto
LEE, G. M. ditto ditto

MACDONALD, General G., Torquay, Devon.
MACAUSLAND, J. C. Esq., The Lusties, Co. Fermanagh.
MANLY,* Major J. S., late 27th Inniskillings.
MAUDE, Colonel, C.B., Judge-Advocate-General, Bombay.
MAUDE, Maurice C., Esq., J.P., Lenaghan, Enniskillen.
MONTGOMERY, Captain H. de Fellenburg, late Fermanagh Light Infantry.
MECHAM, J. R., Esq., 26th Cameronians, late 27th Regiment.
MICHAELSON, G. H., Esq., 27th Inniskillings.
MATTHEWS, W., Sergeant-Major, Armagh Militia, late 27th Regiment.
M'CLELLAND, E., Colour-Sergeant ditto ditto
MEARA, J., Sergeant, Roscommon Militia, late 27th Regiment.
MANN, J., Colour-Sergeant, 27th Inniskillings.
MONRO, J., Cook-Sergeant ditto
MEADE, W., Sergeant ditto
M'GAHEY, J., Corporal ditto
MOLES, W. J. ditto ditto
M'COMAS, R., Private ditto
M'GOLDRICH, J. ditto ditto
M'KERVEY, H. T. ditto ditto
M'MANNUS, M. ditto ditto
M'KINLEY, T. ditto ditto
MAGUIRE, J. ditto ditto
MAGUIRE, D. ditto ditto
MAGEE, J., Drummer, Armagh Militia, late 27th Regiment.
MURPHY, Joseph, Private, 27th Inniskillings.
MORRISON, R. ditto ditto
MOORE, T. ditto ditto
MASTERSON, W. ditto ditto

NASH, Thos., Surgeon-Major, late 27th Inniskillings.
NOLAN, J., Orderly-room Clerk, 27th Inniskillings.
NUNAN, M., Master Tailor ditto

* Son of Captain Manly, who served with the regiment throughout the Peninsular war.

NOBLE, T., Sergeant, 27th Inniskillings.
NOBLE, A., Corporal ditto
NORRIS, J., Private ditto

O'DONNELL, Lieutenant-Colonel J. V., late 27th Inniskillings.
O'CALLAGHAN, A., Esq., B.A., Londonderry.
O'LEARY, Rev. David, M.A., Chaplain to the regiment at Enniskillen.
Officers' Mess, 27th Inniskillings.
OSBORNE, J., Private, 27th Inniskillings.
O'NEIL, J. ditto ditto

PICKERING,* Mrs. A. H., 24, Kensington Gate, London.
PATTON, Captain H. B., late 27th Inniskillings.
PHILLIPPS, Captain H. B. P., 27th Inniskillings.
POTT, Captain W. ditto
PINWILL, W. Stackhouse C., Esq., late 27th Inniskillings.
PLUNKETT, Thomas, Esq., T.C., Enniskillen.
PRICE, Alex., Esq. ditto
PRENTICE, J., Sergeant, 27th Inniskillings.
PAGE, J. N. ditto ditto
PEVERETT, A. ditto ditto
PETTY, T., Corporal ditto
PROCTOR, J., Private ditto
PULHAM, G. ditto ditto
PARKER, F. ditto ditto

RUMLEY, Lieutenant-General Randal, C.B., Head Colonel 27th Inniskillings.
Reading-Room, 27th Inniskillings.
Reading-Room, Lisnaskea.
ROBINSON, J. R., Sergeant, 27th Inniskillings.
REEVE, W. ditto ditto
ROBINSON, R. ditto ditto
ROWE, W., Corporal ditto
RANKIN, T., Private ditto
REID, T. ditto ditto

SMITH, Lieutenant-General Thomas Charlton, late 27th Inniskillings.
STEELE, Rev. W., D.D., Portora, Enniskillen.
STAINFORTH, P., Esq., 27th Inniskillings.
Sergeants' Mess, 6th Inniskillings.
ditto 27th Inniskillings.
SLOAN, G., Colour-Sergeant, 27th Inniskillings.
SMITH, J., Private ditto

* Daughter of General Reeves, Colonel 2nd-27th in Spain.

Seaman, J., Private, 27th Inniskillings.
Simpson, W. ditto ditto
Stutters, H. ditto ditto
Stonham, D. ditto ditto
Scollan, T. ditto ditto
Smith, P. ditto ditto
Sparks, Jas., Private (Band) ditto

Thesiger, Colonel Hon., commanding 6th Inniskillings.
Tottenham, A. Loftus, Esq., D.L., J.P.
Teevan, Dr. Thomas,* Raceview, Enniskillen.
Trenor, Captain J., late 27th Inniskillings.
Taylor, Captain D. M., 27th Inniskillings.
Tunnard, H. S., Esq. ditto
Tummoney, J., Colour-Sergeant, ditto
Thompson, W., Private ditto
Taylor, E. ditto ditto
Taylor, J. ditto ditto

Verschoyle, Rev. Canon R., M.A., Derryvullen Rectory.

Williamson, Major-General, Mallow, late 27th Inniskillings.
Wood, John A., Esq., J.P., Hollyhill, Enniskillen.
White, Captain R. W. E., 27th Inniskillings.
Wodehouse, H., Esq., Adjutant ditto
Williams, H. P., Esq. ditto
Werner, Louis, Bandmaster ditto
Windrum, W., Sergeant-Major ditto
Withers, W., Sergeant-Major, Dublin City Militia, late 27th lings.
Windrum, W., Hospital Qr.-Master Sergeant, late 27th Inniskill
Windrum, W., Drum-Major, 27th Inniskillings.
Windrum, J., Colour-Sergeant ditto
Williams, W. ditto ditto
Windrum, J., Corporal ditto
White, H. ditto ditto
Watts, C., Private ditto

Young, A. H., Esq., 27th Inniskillings.

* Nephew of Surgeon Teevan, sometime attached to the 2 killings.

THE 27TH INNISKILLINGS

"The regiment . . . . is, and always has been, *unsurpassed* in glorious action, and in good conduct *unsurpassable*."—Lieutenant-General Sir W. Napier, K.C.B

THE TWENTY-SEVENTH,

OR

INNISKILLING REGIMENT OF FOOT,

BEARS ON ITS COLOURS

A CASTLE WITH THREE TURRETS, ST. GEORGE'S COLOURS FLYING, WITH THE WORD "INNISKILLING" ABOVE, AND THE NUMERALS "XXVII" BELOW; ENCIRCLED BY A WREATH OF THE ROSE, SHAMROCK, AND THISTLE, AND SURMOUNTED BY A CROWN; IN COMMEMORATION OF THE GALLANT DEFENCE OF CROM BY THE VOLUNTEER CORPS IN 1689;

THE WHITE HORSE, WITH THE WORDS "NEC ASPERA TERRENT," FOR THE REGIMENT'S SERVICES IN HANOVER;

THE WORDS "ST. LUCIA," TO DENOTE THE GALLANT CONDUCT OF THE REGIMENT IN THE ATTACK ON MORNE FORTUNÉ IN 1796, WHICH LED TO THE SURRENDER OF THAT ISLAND;

THE SPHINX, WITH THE WORD "EGYPT," TO COMMEMORATE THE SERVICES OF THE FIRST BATTALION AT THE BLOCKADE OF ALEXANDRIA, 1801;

THE WORD "MAIDA," IN MEMORY OF THE "FIRMNESS" OF THE FIRST BATTALION IN THAT BATTLE ON JULY 4TH, 1806.

ALSO THE WORDS "BADAJOZ," "SALAMANCA," "VITTORIA," "PYRENEES," "NIVELLE," "ORTHES," "TOULOUSE," AND "PENINSULA," IN TESTIMONY OF THE DISTINGUISHED SERVICES OF THE THIRD BATTALION IN THE SEVERAL ACTIONS DURING THE WAR IN PORTUGAL AND SPAIN AND THE SOUTH OF FRANCE, FROM 1809 TILL 1814;

AND THE WORD "WATERLOO," TO DENOTE THE UNDAUNTED BRAVERY OF THE FIRST BATTALION IN THAT BATTLE ON THE 18TH OF JUNE, 1815, NEARLY TWO-THIRDS OF ITS NUMBERS HAVING FALLEN KILLED OR WOUNDED IN ITS SQUARE.

THE HISTORICAL RECORD

OF THE

27TH INNISKILLING REGIMENT.

INTRODUCTION.

ENNISKILLEN, the county town of Fermanagh, became celebrated by reason of the heroism of its inhabitants during the revolutionary period of 1688 and 1689, when, like the men of Derry, they made a decided stand in favour of civil and religious liberty. Not only is Enniskillen celebrated in military story, but it is unique in one respect—it is the only town in the three kingdoms which has given its name to two regiments; and those two regiments, at all times in great measure supplied by men from Fermanagh, have become celebrated by their valour and intrepidity, worthy followers of the great men who rendered the name of their island town conspicuous in history. The 6th Inniskilling Dragoons have not for over twenty years been quartered in Enniskillen, owing to the insufficiency of barrack accommodation for cavalry, but at the present time two troops lie twenty miles off, at Belturbet (head-quarters, under Colonel Thesiger, at Dundalk), while the head-quarters of the 27th Inniskilling Foot, under Colonel Freer, at present lie in our Main Barrack.

There is one matter of regret connected with the regimental Records—that they are particularly hazy about 1688 and

1689. I am given to understand that they were more full than at present at one time, but the wish to avoid any matters that might affect the feelings of susceptible people resulted in a brief paragraph being considered sufficient for the purpose. I am happy to say that I shall be able to give full details of this important epoch in the regiment's history, and I hope in a spirit free from animus or party feeling, for the incidents of that date still provoke controversy. There is another matter of regret—that a wide gap remains in the Records between 1690 and 1736, one which I am only in part able to fill. This is the more lamentable, as the intervening period was very eventful—ripe with stirring incidents, when the aptly termed "Battle-Field of Modern Europe," Spain and the Low Countries, was the chief theatre of a war in which Gibraltar was taken by the British (July, 1704), and in which the genius of Marlborough humbled the power of France. At Blenheim in Bavaria, at Ramillies in South Brabant, at Oudenarde in East Flanders, and at Malplaquet on the north-eastern frontier of France, the ambitious Louis XIV. was driven to negotiate for peace, till the Treaty of Utrecht gave rest to exhausted Europe.

AFFAIRS IN INNISKILLING.

The evident designs of King James II. to subvert the Protestant institutions of England, and the complete restoration of the Romish worship, and his unconstitutional acts to effect that purpose, spread consternation among his Protestant subjects in Ireland, as well as in the sister country. The recall of the Duke of Ormond to London, and the appointment of Tyrconnell as lord-lieutenant, contributed to their uneasiness, as his bigotry and prejudices for the Catholic cause were well known. Rumours of another massacre of the Protestants and the attitude of the Irish Catholics increased the alarm, which was largely shared by the Protestants of the little town of Inniskilling [modernized Enniskillen]. A copy of the anonymous letter to Lord Mount-Alexander, announcing the intended massacre of

Protestants, reached them on the 7th of December—the day that Derry closed its gates against the Redshanks; and although there, as elsewhere, the 9th passed over quietly, the popular alarm excited by the letter did not pass away. On the 11th a letter was received from the Government authority in Dublin, directing them to make arrangements for having two companies of infantry quartered in the town. This doubled their uneasiness.*

The people were in perplexity as to what in these circumstances ought to be done. To assume an attitude of resistance to the constituted authorities of the country was no light matter. On the other hand, rumours of a massacre were rife. The native Irish in the neighbourhood were providing themselves with arms; it was an unusual thing to have a garrison planted amongst them; and the probability as they believed was, that the day for cutting their throats was only postponed until everything was ready, and till, with the assistance of the soldiery, it could be done with the greater safety and convenience. While the town was in this state of uncertainty as to what might be done, five men came together and resolved to refuse admittance to the soldiers, whatever consequences might ensue. The Prince of Orange, as they knew, had landed in England some five weeks before; civil war was imminent in Ireland; North and South most likely would be pitted against each other; and it appeared to them that by refusing to admit the troops they might be able, not only to protect themselves, but to hold the most important town between Connaught and Ulster in the interest of their party. [The important military situation of Enniskillen is not now lost sight of by the authorities.] However plausible such considerations, it was nevertheless a mad resolve, in face of the facts; which simply were, that arrayed against them was the whole power of the Irish Government, and that all the means of resistance Inniskilling had were ten pounds of powder, twenty firelocks, and eighty men. The five men, however, did resolve; sent notice of their determination to the surrounding country, and craved its assistance; set carpenters to work at the drawbridge in

* Professor Witherow.

connection with the stone bridge lately erected at the east end of the town, and, like men in earnest, took every step that they could think of to increase their power of resistance.

INNISKILLING IN 1688.

It should be recollected that the East Bridge was at that time only half its present width. Inniskilling was altogether different from the Enniskillen of to-day. The castle—now incorporated with the barracks of that name—was surrounded by a canal. A ditch ran along what is now the Main Barrack square. The only places where houses existed were along the present Main Street, Eden Street, Water Street, and Market Street. The barracks, capable of holding two companies of foot, occupied ground in the present High Street. The sessions house and jail were where the court-house now stands. The meeting-house (Presbyterian) was situated as at present. The Royal Free School-ground was about Mr. Innes' premises in Market Street. The corn market was then a common extending from the East Bridge, and the present jail occupies the ground of the then Gallows Green. From the church to the present Main Barrack point was green pasture, and so on round the island.

DEFENCE RESOLVED UPON.

We left the men of Inniskilling in rebellion. Perplexity prevailed. Captain Corry, the ancestor of the present worthy owner of Castlecoole, the Earl of Belmore, was averse to refusing the king's soldiers admittance. The provost, Paul Dane, intended to give the soldiers entrance, but irresolution still prevailed. Gustavus Hamilton gave his influence to the side of those who thought the place should be defended, the drawbridge was completed, all the Roman Catholics residing in the place were sent away, and the Protestants of the surrounding country invited to come in and assist in the defence.

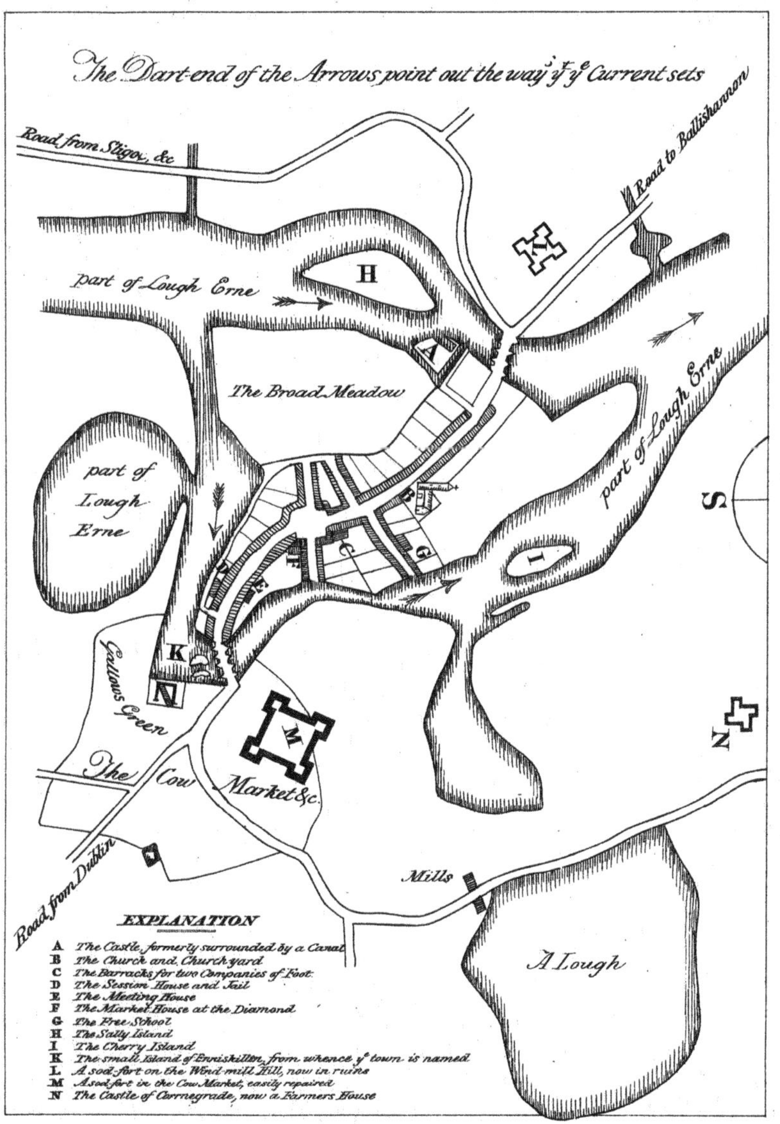

MAP OF ENNISKILLEN IN 1689, FROM HARRIS'S LIFE OF WILLIAM III.

Marcus Ward & Co.

INSTITUTION OF THE REGIMENT.

Gustavus Hamilton was elected governor of the town on the 15th of December. The Cole family, of which the Earl of Enniskillen is now the head, came to Ireland among the colonists of James I., and settled in Fermanagh in 1611. The Sir Michael Cole of this time was then in England; and Mr. Hamilton, who had lived five miles out of town, took up his residence in Sir Michael's residence—the Castle of Inniskilling. The governor organized the townsmen into two companies of foot. One of these was under Captain Allen Cathcart; the other, under Captain Malcolm Cathcart, consisted mostly of Presbyterians — "that party," says MacCormick, "effectually espousing our interest, and never declined us in the most dangerous times." These two companies were the foundation of the 27th Regiment of Foot.

It might not be uninteresting to know the obligation of the private soldiers of those two companies. It was as follows:—

"I, A.B., do hereby testify and declare, and upon the Holy Evangelists, swear, that I will own and acknowledge Gustavus Hamilton, Esq., Chief Governor of this town of Inniskilling, and shall give due obedience to him and my superior officers in all his and their commands, and shall to the utmost of my power and ability defend him, them, and this place, with the country adjacent, together with the Protestant religion and interests, with my life and fortune, against all that shall endeavour to subvert the same. So help me God, and the holy contents of this book."

ROUT OF LISMELLA.

The Inniskilling men sent to Derry for arms and ammunition, and on the next day, 16th December, news came that the two foot companies sent by Tyrconnell, then lord-lieutenant, had reached Lismella [modern Lisbellaw]. "Most of the townsmen were engaged in public worship at the time, but soon retired, took up their arms, and put themselves in array. Notwithstanding all the help sent them by the country, their whole strength did not exceed 200 foot and 150 horse ill armed, and with *no military* training or *experience.* They left town with the intention of persuading, if possible, the soldiers to return, but prepared, if necessary, to resist their entrance. Rumour magnified alike their numbers and determination. No sooner did the soldiers come in view of the Inniskilling men, than, without waiting for their approach, they turned and fled to Maguiresbridge, whither they were followed by their officers, who, at the time when the encounter was imminent, were dining quietly at Captain Corry's, not dreaming of an armed resistance to Government orders. Next day they fell back to Cavan, where they awaited the commands of Tyrconnell."

DEFENCE OF INNISKILLING.

The Inniskilling delegates from Derry returned with the assurance that Derry was willing to help them. On their way home they presented a letter explanatory of their conduct to Lord Mountjoy, who told them they must admit the king's soldiers—there was no help for it. About the end of December, 1688, a letter was received from Lord Blaney, advising them to continue their refusal to admit the king's troops. The great portion of the next month, January, 1689, was spent in getting ammunition, bettering their position, enrolling new men in the regiments, and preparing for a siege.

On the 23rd of this month the town sent a letter to Lord Mount-Alexander, of the Protestant party; and in it the people of Inniskilling made a declaration, by which they faithfully abided—"WE STAND UPON OUR GUARD, AND DO RESOLVE BY THE BLESSING OF GOD RATHER TO MEET OUR DANGER THAN EXPECT IT." A nice answer was received, and information that Lord Mount-Alexander had been appointed commander-in-chief of the Protestant forces in the north-east of Ulster.

The (afterwards 27th) regiment of foot had now swelled to ten companies of 72 private men each, and a troop of 100 horse, well armed with carbines and pistols. A declaration was published on 27th January, that having regard to the great and imminent danger of a massacre, that the Protestants of the county and country adjacent do assemble in Inniskilling with arms and provisions for their defence. A large meeting was then held in order to make a show of strength, to give waverers an opportunity of choosing their side, at which Governor Hamilton was chosen colonel in command, and Thomas Lloyd lieutenant-colonel.

ALLEGIANCE TO WILLIAM AND MARY.

Early in March news arrived that James had deserted the throne, and that William III. and Mary were proclaimed, whereupon the inhabitants of Inniskilling pledged their allegiance to the new king and queen. Colonel Lundy, of Derry, ordered the people of Cavan and Inniskilling to fall back on Derry. The latter declined, believing it best to defend their own town, and thus divide the enemy's forces. To the astonishment of Inniskilling, the better-class people of Cavan, with horse and foot companies, entered their town for rest, and after three days' stay, and vainly imploring the men of Inniskilling to go with them, departed. Lord Galmoy, at the head of a Jacobite army, was drawing near and had reached Lisnaskea. He sent forward a message to Inniskilling to surrender. Professor Witherow says:—"The Enniskilleners of course meant to fight: it never occurred to them that they could do otherwise: and they used every argument they could think of to induce the Cavan men to stay and assist. But it was in vain—they would and they did march on to Derry. Seeing that they were bent upon this, the governor proclaimed that all the men from Cavan on the way to Derry must take their wives and children with them, else, if left behind, they would be turned immediately out of town. This had a good effect. Some three or four companies of foot, who could not conveniently take their wives and children with them, fortunately for themselves, were obliged to stay; but all the rest, both horse and foot, and even the officers of some of the companies that remained, pushed on to Derry. They would have persuaded the people of Enniskillen to go with them; but the Enniskilleners to a man determined to stay and to defend a position which they rightly regarded as the key of Ulster on the side of Connaught, and which if they were to lose, Derry itself could not be maintained against the accession of strength thus brought to the assailants. They turned away alike from the entreaties of Cavan, the orders of Lundy, and the summons of Galmoy. The latter they answered by sending a letter to say, that they held their town in the interests of King William and Queen Mary, and would defend it to the last. Their attitude was always in the words of their gallant governor—'*We stand upon our guard, and do resolve by the blessing of God rather to meet our danger than to expect it.*'"

We have now arrived at the time when the actions of the men of Inniskilling are to be related—the actions of those two regiments which have made their town famous, and given glory to their first soldiers; for whether a man be Catholic or Protestant, he must recognise the heroism, valour, and intrepidity of the men, just as we must equally respect the defenders of Derry and Limerick.

FIRST SIEGE OF CROM.

The most advanced outpost of the Inniskilling camp in Dublin direction was placed at Crom Castle, about thirteen miles away, and this castle Lord Galmoy resolved to take. The sequel is thus told in "Derry and Enniskillen:"—

"His lordship had no cannon to enable him to capture any place of strength; but he hoped to make up for his lack of artillery by an invention of his own. He made two cannon of tin, bound them round with whip-cord, covered with buckram, so as to give them the appearance and colour of real guns, and had them drawn in the direction of Crom with great noise and apparent difficulty by eight horses apiece. So soon as he had his mock guns planted in position, he threatened to batter down the castle if it was not immediately surrendered. But the little garrison was not so much intimidated as he had expected: their only answer to his threat was a volley of fire-arms. They knew that if it was at all possible Enniskillen would send them relief. Nor were they mistaken in their hope. On Saturday, 23rd of March, on the day that the Cavan men left for Derry, Enniskillen drew out all its horse and foot to measure its strength with Galmoy, who had now advanced to Lisnaskea, within ten miles of the town. The enemy did not make his appearance that day; but in the evening the intelligence arrived that Galmoy and his men were engaged in the siege of Crom. General Hamilton decided to send relief to the little garrison. Choosing out 200 of his best armed men, he sent some of them in boats and some by land, in hope that they would reach Crom and get into the castle before daylight; but the day had already dawned before they succeeded in reaching the place of their destination. The besiegers were aware of their arrival, and fired at them; but their aim was so ill-directed, that they killed only one of the boatmen, and failed in preventing the relieving party from entering the castle and uniting with the garrison. The united body then sallied out of the fortress, attacked Galmoy and his men with the greatest violence, beat them out of the trenches, killed thirty or forty, captured the two tin guns, and compelled him to raise the siege and to retire to Belturbet."

The Captain Crichton then owner of Crom, was ancestor of the present representative of the family, the Right Hon. the Earl of Erne, K.P.

LUNDY AND INNISKILLING.

On the 25th of March news came that a supply of arms and ammunition had arrived at Derry; but all the men of Inniskilling could get from Colonel Lundy was five barrels of powder and sixty old musket barrels, which had been "cast" as lumber. These were utilized by the good men of Inniskilling; and this was their only war material, except what they took from the enemy, till their relief by Major-General Kirke. Lundy again invited the Inniskilliners to vacate their town this time, and to maintain the fords of the Finn. They were not regular soldiers, however, bound to obey his command, and so declined, on their own judgment. Their wisdom was fully verified when they became aware of his traitorous conduct at Derry, and the deception practised on Sligo and Cavan. Before leaving for England, Lord Kingston sent to Inniskilling those of his forces not required to garrison Ballyshannon and Donegal. The two troops of horse and the six companies of foot thus acquired added much to the strength of the garrison. Thus did the (afterwards) 27th Foot and the 6th Dragoons become formidable regiments.

TRILLICK AND AUGHER.

Having heard that the Jacobites were about to plant a garrison at Trillick, nine miles from Inniskilling, on the Derry road, Colonel Lloyd, on the 24th April, led out a party to dislodge them. His approach was discovered so early that the enemy had time to escape before he came up; a hot pursuit of six hours dispersed them over the country without enabling Lloyd and his men to overtake them, but they brought back with them both the enemy's baggage, and a large quantity of cattle and provisions, which were taken from the surrounding neighbourhood.*

Some days after, hearing that the enemy were about to plant a garrison in the Castle of Augher, which lay eighteen miles from Inniskilling on the road to Charlemont, Colonel Lloyd and his men marched, in the hope of being able to surprise them, on the morning of Sunday, the 28th of April; but, notwithstanding the rapidity of their march, the garrison had notice of their approach, and fled, taking with them all that they could carry. The Inniskilliners had to

* Professor Withcrow's account, whose compendium is most accurate.

content themselves with burning the castle, levelling the fortifications, and seizing as many cattle as the neighbourhood could supply. They then passed through the mountains, with the view of expelling the garrison that the enemy had placed in the house of Daniel Eccles, Esq., of Clones; but that garrison also had notice of their approach, and, of course, set fire to the building, and took flight before they came forward. Having swept across a large part of the counties of Monaghan and Cavan, they returned home on Thursday, 2nd May, bringing with them horses, sheep, cattle, and provisions in great abundance. The plan adopted at Inniskilling from the first, and acted on to the last, was, to go out and fight every enemy before he came near the town. The result was, throughout the whole campaign, no enemy got leave to come within sight of that place, and that, while Derry was living on starch and tallow, Inniskilling never knew what it was to want. Often during the war a good milch cow could be bought for eighteenpence, and a cow not giving milk for sixpence.

ROUT OF BELLEEK.

Two days afterwards Captain Folliott, governor of the Protestant garrison at Ballyshannon, asked help, as he was besieged by the Connaught Jacobites. On the 6th May Lloyd started for Ballyshannon. The Jacobites heard of his approach, and marched as far as Belleek to meet him, where they placed themselves in an advantageous position in a narrow pass, with the lough on the one side and a bog on the other. Lloyd succeeded in flanking them by means of a path to the right not much known, and the enemy turned and fled, without firing a shot. Not less than 190 of the horse were slain in the pursuit, but most of the foot, escaping through a bog, made good their retreat to Sligo. The Inniskilliners pushed forward and took 60 of their men prisoners on the Fish Island at Ballyshannon. The siege was then raised, and the total injury of the Inniskilliners was one man wounded.

INNISKILLING FORT.

During all this time the town was carefully guarded, provisions were kept in store, all boats on the lake were brought to Inniskilling to prevent surprise by water, cannon and fire-arms were kept fit for service, and a horse patrol went out every night to scour the country, so as to prevent any sudden attack. In addition, the construction of a fort was commenced on a hill overlooking the East Bridge, which was not finished till June, and, when completed, it was joined to the East Bridge by a line of communication. The fort still exists, and the site is known to this day by the name Forthill, whereon the monument of an old colonel of the 27th is erected—General Cole. Its commanding position for military purposes rendered it of great utility; and with the fort on the other or west side, where the redoubt or hospital now stands, and a belt of water, Inniskilling was well protected.

RAID INTO CAVAN AND MEATH.

At the end of May the Inniskilliners organized another expedition for the purpose of checking the Jacobites. Their fame travelled before them. At Wattle Bridge the enemy retreated, and at Redhill the garrison surrendered. Ballinacarrig Castle also surrendered, by which they obtained a large supply of arms and ammunition and rich provisions. Having scoured the country on to Kells in Co. Meath, *within thirty miles of Dublin*, and having struck terror into the metropolis itself, where their strength was believed to be much greater than it really was, they returned to Inniskilling *without the loss of a single man*, bringing back with them about 3000 cows and oxen, 2000 sheep, and 5000 horses, laden with meal, wheat, and malt. The following testimony is born to their conduct by one of their enemies:— "They were the fairest enemy that ever came into a country, not injuring any person that lived peaceably, leaving a troop of horse in the town of Cavan until all the army were marched away, to see that no injury might be done to the common people."

What brave fellows these must have been, and what a contrast to the cut-throat Carlist insurgents of the latter half of the nineteenth century, the boasted age of civilization and progress! All this time, while they had cattle in plenty, Derry was in a state of starvation. Yet they were helping Derry by keeping off more foes. Why not attack the capital? asks Witherow. A bold attack on Derry when most of the army were engaged at Derry might have been successful; but for the Inniskilliners to attempt it then would have been a fatal mistake, seeing that they had no large guns, that their supply of any kind of arms and ammunition was very scanty, and that at that time, when they were absent on this raid, Sarsfield, at the head of 6000 or 7000 men, was lying in the park of Sir William Gore, near Manorhamilton, about sixteen or seventeen miles from Inniskilling.

CAPTURE AT OMAGH.

On Monday, 3rd June, two foot companies, under Captain Henry Smith and Captain Robert Corry, marched to Omagh, and took away 150 troop-horses from the Jacobite garrison there. By this action they disabled three troops of the enemy and mounted three troops of their own much better than before. The garrison had time to save the town, but not the horses, from capture.

FAILURE TO RELIEVE DERRY.

Soon after news came of the distress in Derry. Governor Hamilton, knowing that if Derry were to fall he could not hold out against the full force of King James's army, which could then be directed against Inniskilling, determined to make an effort to raise the siege, or failing that, to throw supplies into the town. The men would have wished their usual commander, Lloyd, to be at their head, but the governor was resolute in his determination. As he approached Omagh the enemy retreated into the town, and fortified themselves in the house of Captain Mervin, at the end of the town. The garrison refused to surrender. Before anything could be done, it was ascertained that Lord Clancarty was on his way to reinforce the Jacobite army at Derry, and would be at Omagh in a few hours.

Seeing that they had no provisions for themselves at Derry, and that the country had been wasted, they considered it would be foolhardy to place themselves in a dangerous position between Clancarty and James's army at Derry, especially as Sarsfield might take the opportunity of attacking Inniskilling. Besides, they had only 1500 men, two cannon, and inferior arms. In such circumstances it would have been madness to have persevered in their original design, which was to march down the eastern bank of the Foyle, to fall on the detachment at Waterside, which was separated by the main body of the besiegers by the river, and, having conveyed provisions into the city, to make good their retreat homewards through the Monteloney mountains. Had Colonel Lloyd been at their head, they probably would have measured their strength with Clancarty, who was then approaching from Dungannon, but they did what in their circumstances was, perhaps, the best thing they could have done: they returned immediately home. Out of respect to the proprietor, Captain Mervin, they did not burn Omagh; but they might as well have done so, for it was burned a few weeks afterwards by the Jacobite army, on its retreat from Derry.*

BATTLE OF BELTURBET.

Shortly after the failure to relieve Derry, news came from Colonel Crichton, of Crom, and Captain Wishart, who commanded at one of the outposts, that a strong party of the

* Hamilton and MacCormick.

enemy had advanced as far as Belturbet, with the intention, it was believed, of taking Inniskilling. On the very night of the receipt of the intelligence, Colonel Lloyd drew out all his men, and next day reached Maguiresbridge. The numbers of the Inniskilling men were as usual magnified. Sutherland heard of their approach, and their number being fixed at 15,000, he at once retreated with his main body to Monaghan, leaving Colonel Scot with 80 dragoons and 200 foot to hold the church and the churchyard, the only place of strength in the village. Lloyd and his officers thinking it vain to follow Sutherland, on Wednesday, 19th June, they marched on Belturbet; and within two miles of that village, the dragoons of both parties came in sight of one another. The horse of the royal forces were driven back to Belturbet, and the Inniskilling men so galled the garrison in the church, that at the end of two hours it surrendered; three hundred prisoners were taken, two barrels of powder, seven hundred muskets, fifty-three dragoon horses, and as many red coats as served for two companies. In addition, a great quantity of the provisions, amounting to twenty tons of bread, flour, wheat, and malt, was sent to Inniskilling by water. Thirteen commissioned officers were taken prisoners. The two hundred common soldiers were taken to Inniskilling, and were employed at the finishing of the erection of the fort near the East Bridge (now the Forthill).

Early in July, thirty barrels of powder were received from the *Bonadventure* frigate at Killybegs, sent round for the purpose of giving assistance by Major-General Kirke, then lying in Lough Swilly.

DEFEAT AT CORNAGRADE.

The heavy blows dealt the Jacobites by the frequent excursions of the men of Inniskilling caused the Duke of Berwick (natural son of King James) to be detached from the main body before Derry, with a flying division. It scoured the country on all sides. Having visited Donegal,

Berwick formed a junction with Sutherland; they encamped at Trillick for a few days, and then advanced on Inniskilling. Colonel Lloyd at this time was on a visit to the *Bonadventure* frigate. Governor Hamilton ordered two troops of horse under Captain Hugh Montgomery and Captain Francis King, in company with 100 foot, to advance as far as Kilmacormick hill, and there to engage the enemy, promising to immediately send supports.

[This battle is more familiarly known as the battle of Cornagrade, though the title is rather incorrect. The present old road divides the hill of Kilmacormick and nearly the townland from Cornagrade. Cornagrade overlooks the present Main Barrack and the west end of the town.

The reinforcements did not come to the Inniskilling men in time. The horse of the enemy were driven back by the foot soldiers, but the Inniskilling horse fled from the ground, leaving MacCarmick and his foot soldiers to their fate. The little party were then surrounded; some of them were slain, and the rest made prisoners. This was the first defeat sustained by the Inniskillings.

It was at this action an incident of "bravery and determination" occurred which has made the name immortal, and which is worthy of record beside the gallantry of Captain Waldron, of the 27th, in the Peninsula. Hamilton tells the story in his "Actions of the Inniskilling Men;" and Harris, in his life of King William, relates it thus:—

"John Wilson, a foot soldier, in this general slaughter of his companions, stood the shock of several of the troopers, who, all together, were hewing at him. Some he stabbed with his bayonet, others he knocked down with his musket, and when his arms dropped from his hands he leaped up at them, tore down some, and threw them under their horses' feet. At length, oppressed with twelve desperate wounds, one of which was quite across his face, so that his nose and cheeks hung over his chin, he sunk down in a scrubby bush. While he was bleeding in this condition, a sergeant darted his halbert at him with such fury, that he struck it through his thigh, and could not draw it out again. Wilson, roused as from death, made his last effort, tore the halbert out of his thigh, and, collecting his whole strength, darted it through the heart of his enemy. By the assistance of the halbert he dragged his mangled limbs to Enniskillen, where he was wonderfully cured, and lived thirty years after."

The Duke of Berwick, being a gentleman as well as an officer, treated his prisoners kindly, and gave orders that no man, on pain of death, should rob them of their property. For some reason or other, perhaps because he had no weighty guns, and because he knew that the town was protected by the cannon of the fort, he did not attack Inniskilling. He withdrew his troops immediately after the action, as if content with what he had gained, and scoured the country from Omagh to Ramullan, up till the time that the siege of Derry was raised.

THE REGIMENT COMMISSIONED.

Some commissioners despatched to Kirke in Lough Foyle, from Inniskilling, received fresh supplies, with commissions for two regiments of horse and three of foot. This was the first commission of the 27th. There also came seven or eight of his best officers, under charge of Colonel Wolseley. They came home by way of Ballyshannon, Belleek, and then by water to Inniskilling, where the whole garrison turned out to meet them. On that very day, while the people of Inniskilling were rejoicing in help from England, the *Mountjoy* broke the boom at Derry.

The following is a copy of King William's warrant authorizing the formation of the regiment:—

"William R.

"Whereas We have thought fit to forme a regiment of horse, together with 2 regiments of dragoons, and 3 regiments of foot out of Our Inniskilling forces, and to take them into Our pay and entertainment, We do hereby pass this Our establishment of the said forces, to commence on the first day of January, 1689-90, in the first year of Our reign.

	Officers and soldiers.		Amount per annum. £.	s.	d.
"A regiment of horse * of 12 troops ...	714	...	40,207	15	7
Two regts. of dragoons of 8 troops each	1162	...	41,415	6	8
Three regts. of foot of 13 companies each	2781	...	48,435	10	0"

LISNASKEA.

On the evening of the day that saw Derry relieved, word came from Crom for help, that General Macarthy (Lord Mountcashel) was besieging the castle. Next day a similar message came. Governor Hamilton being now ill of fever, Colonel Wolseley took the chief command, and sent to Ballyshannon for as many men as could be spared. Notwithstanding that Sarsfield lay at Tullaghan, beyond Bundoran, the gallant garrison sent 400 or 500 men. After their long march of twenty miles on Tuesday evening, when they arrived at Inniskilling, instead of being fatigued, they declared themselves ready to go on that very night. On the Monday previous, Lieutenant-Colonel Berry had been despatched to garrison Lisnaskea before the enemy could take it, and to ascertain the strength of the Jacobites. When he reached Lisnaskea, he found the castle and the place of so little importance, that he camped all night on the open ground. On Wednesday he marched on Macarthy, and at Donagh the enemy came in sight. Berry immediately retired. He ascertained the enemy was double his strength, and sent off an express to Colonel Wolseley for help. Berry retreated through Lisnaskea in good order. When about a

* The regiment of horse was disbanded after the Treaty of Ryswick. The two regiments of dragoons were retained on the establishment of the army, and were subsequently numbered as—

The V. Royal First Dragoons,

The VI. or the Inniskilling Dragoons.

The three regiments of foot were consolidated into one, which is now the 27th, or the Inniskilling Regiment of Foot. The Londonderry forces, incorporated at the same time, were afterwards disbanded.

mile from the town on the Inniskilling side, he halted at the remote end of a narrow pass which ran through a bog. Here he determined to wait till relief came, and placed his forces in order. Scarcely was this accomplished when the enemy came in sight, commanded by Colonel A. Hamilton. Cross-fire immediately ensued. Colonel Hamilton was wounded, the officer second in command was shot dead, and in the confusion, an ambush party of the Inniskilling men opened fire. The Jacobites then turned and fled. Two hundred were slain and thirty taken prisoners.

NEWTOWNBUTLER—31ST JULY.

The foregoing occurred about nine a.m. Two hours later word was received from Wolseley, and the two officers met at the moat on the old road above Lisnaskea. So great had been the haste to relieve Berry, that no provisions had been brought, and it became a necessity to engage the now advancing enemy, Macarthy himself, or retire immediately. The men were consulted, and all agreed unanimously to advance and fight. Altogether they had something under 200 men, while Macarthy had nearly double that number. Macarthy had raised the siege of Crom, and had reached Newtownbutler to meet the Inniskilling forces. About half-a-mile beyond Donagh the advance-guards of the two armies saw each other. The Jacobites returned towards Newtownbutler. They were driven by Wolseley from a position they had taken near that town, to which they set fire as they withdrew. They halted at the remote end of a narrow causeway running through a bog about a mile on the Clones side of Newtownbutler. Their foot was disposed, the horse placed on the hill, and the cannon deposited to command the causeway.

"The Inniskilliners advanced against them in the same order as before; the foot took the bog, and the horsemen kept the road. The cannon of the enemy, however, played so incessantly, that the Inniskilling horse could not advance a single step along the hazardous pathway. As the foot made their way across the bog, the enemy from behind their cover fired briskly at them; but still they went on steadily until they had gained firm ground, and then, dashing forward with force, they beat them out of their shelter, seized the cannon, killed the gunners, and commenced deliberately to ascend the hill. No sooner were the cannon silenced, than the Inniskilling horse dashed over the causeway at full speed, to take their part in the conflict. When the enemy's horse, from the top of the hill, saw that the bog had been crossed and the guns captured, they wheeled about and galloped off in the direction of Wattlebridge, leaving their foot to shift for themselves. The foot kept their ground until they saw that their horse were fled, and that the Inniskilling horse were upon them; then they too broke and fled."

It appears that the reason of the Jacobites fleeing, "without an effort to retrieve the fortune of the day," was the mistake of an officer in the confusion giving the general's order as—"Face right-about and march," instead of "Face to the right."

This was the greatest victory yet obtained by the Williamites in the field of battle. 2000 of the enemy were actually slain, 500 drowned in the lough, and 400 were taken prisoners, most of whom were officers, among whom was Lieutenant-General Macarthy himself (Lord Mountcashel). His life was spared, as he had on one occasion saved the life of Colonel Crichton, of Crom. He was not very closely guarded, and subsequently escaped, at the end of five months' captivity, by bribing one of the sentinels.

Sarsfield, having heard of the defeat of his party at Newtownbutler, withdrew to Sligo, and on the 4th of August news came to Inniskilling that James had left Derry.

SLIGO.

Story, in his "Impartial History of the Affairs of Ireland," frequently refers to the "Inniskillingers;" but very frequently speaking of them generally, the reader is unable to distinguish whether the horse or foot are referred to.

Their several actions, however, bear the same traces of valour and intrepidity which have ever marked their conduct. The following paragraph from Story will be found interesting:—

"On Friday, the 27th September, we had news that two days before Collonel Lloyd, with about 1000 Iniskilliners, had defeated a body of the Irish that were going towards Sligo (consisting of about 5000), and had killed 700 of them, taken O'Kelly, their commander, and 40 more officers prisoners, with a great booty of about 8000 cattle, with the loss only of 14 men; upon which news, the General ordered all the Iniskillin horse and foot that were in the camp to draw out, and complemented them so far as to ride all along their line with his hat off; then he ordered the Dutch Guards and the Iniskillin foot to draw into a line to the right of our works, at the west end of the town, where they made three running fires, which were answered by the Iniskillen horse from their camp, and by the great guns upon our works, as also from our ships that lay in the mouth of the river. The enemy admired what all this rejoycing should be for, and were in some trouble at first, suspecting we had got some extraordinary news from England; or that there was an army landed in the west of Ireland (which they themselves must have known before us); but when they understood the occasion, they were not much concerned."

This honorary ceremony occurred before Dundalk, where the "Iniskillin" foot was commanded by "Collonel Teffins."

BOYNE.

The Inniskilling foot were at the battle of the Boyne next year. Their part in the action was not very remarkable. The Inniskilling horse were very prominent, and made several brilliant charges. Suffice it to say that the Inniskilling foot sustained at the Boyne the cause they had so bravely defended, with equal valour. It was all needed, for the Irish Jacobite troops fought as Irishmen ever do, but with a desperation and a gallantry worthy of a better cause.

Story observes: "The Inniskilliners and French, too, both horse and foot, did good service."

It was this body of men, so renowned for their loyalty and bravery, who laid the foundation of the character of the 27th or Enniskillen Regiment, a character which it has never on any occasion failed honourably to maintain, and by a uniform intrepidity and firmness in the field, during a very long course of years and events, and of unremitting employment in every quarter of the globe where the British army has been called into action, to preserve untarnished the fair fame of its ancestors, and to contribute to raise the military character of its native town and county; for although the ranks of the regiment have not been always filled with men from the same part of the country, yet so great has been the pride of corps which animates every soldier belonging to it, it impels him to preserve the reputation and distinguished character of the Enniskillen regiment (whose good name has been so long established), and to transmit it as pure as it descended to him. In commemoration of the brave defence made by the Inniskilliners in the Castle of Crom, when besieged by Lord Mountcashel, the regiment was ordered to bear a badge of a castle on its colours and appointments with the word Inniskilling; and since its formation in 1689, it has been chiefly recruited from Fermanagh and the neighbouring counties.

THE FIRST COLONEL.

After the battle of the Boyne, King William pushed on to Dublin and encamped at Finglass, where, on the 8th and 9th July, his troops passed in review before him. He had appointed Zechariah Tiffin to the colonelcy of the Inniskilling foot on 26th June, 1689,* and I find in Story's "History of the Wars in Ireland," that the number of private men in Tiffin's regiment were numbered at 625.

* Gustavus Hamilton, who had been colonel while the Inniskillings was a volunteer regiment, was removed to the 20th Regiment of Foot, on the 1st June, 1689, in the place of Sir Robert Peyton, resigned, and was afterwards created Viscount Boyne.

SLIGO AND BALLYMORE.

1691.—In March, 1691, Colonel Tiffin made an excursion towards Sligo, and met with no formed body of the enemy, so that his men had leisure to take a good prey, and bring off several families.*

On April 9th, Colonel Tiffin sent 200 foot and 100 dragoons from Ballymore towards Sligo, who in their march killed 42 Rapparees, bringing off a good prey at their return.

On June 6th, Lieutenant-General Ginckel marched from Mullingar to Rathcondra, and were joined upon their march by Colonel Tiffin's regiment. On June 7th Ballymore was attacked, and after two days' siege it surrendered. Athlone was besieged on the 20th, and carried by storm.

AUGHRIM.

On Sunday, the 12th July, was fought the battle of Aughrim. Colonel Tiffin's regiment was placed in the second line in Stuart's brigade. The contest was unusually severe, the Irish defending the ditches "so stoutly." Colonel Tiffin was principally engaged in the bog, where the fighting was hardest. Finally, General Ruth was killed, and his army routed. In this battle the (afterwards 27th) regiment had 19 men killed and 40 wounded.

GALWAY.

Galway was taken in the same month, and among the persons given up by the governor of the town as hostages for the due performance of the treaty, was the then "Lord of Iniskillin."† Colonel Tiffin's regiment was attached to the Prince of Hesse's party, which on the 29th August captured Castleconnell.

* Story. † Lord Maguire.

LIMERICK.

In the attack on Limerick, while a portion of the royal army was crossing the river on 16th September, four or five of the enemy's regiments came out to oppose the landing. Major-General Talmash, with Colonel Tiffin's foot and some dragoons, routed the enemy, taking a French lieutenant-colonel, a captain, and some other prisoners. On September 22nd, Colonel Tiffin's was one of the regiments which crossed the Shannon into the county of Clare; and in the evening, along with Kirk's, St John's, and Hamilton's regiments, attacked the enemy's works that covered Thomondbridge, where the enemy had posted about 800 men. Notwithstanding the fortifications and other advantages, the enemy was routed, with 600 killed. The famous treaty was afterwards drawn up and ratified, the breaking of which has secured for Limerick the title of "The City of the Violated Treaty." Colonel Tiffin's was one of the regiments which took possession of the town on the 4th of October. Some of the Irish regiments which had intended going to France, hearing the bad treatment their brethren who had been removed in the first transports had received, refused to go, and, delivering up their arms to Colonel Tiffin in the month of November, went to their homes.

FLANDERS.

It is much to be regretted that no authoritative information can be obtained from this period until 1735. It will hereafter appear that the actions of the regiment during a portion of this time have been supplied from private sources. The general opinion in the regiment was that the Records of the interval noted above were lost.

In a letter to Mr. Cannon, dated May, 1842, Lieutenant-Colonel Montague Johnston, 27th Regiment, reiterated a former statement—"that all trace of the service of the regiment between 1691 and 1735 had been lost."

But even before that time inquiries were instituted on this subject, as will appear by the following letter:—

"SIR, "St. Vincent, August 20th, 1828.

"I have the honour to acknowledge the receipt of your letter of the 25th June last, signifying to me the commands of the General Commanding-in-Chief, that a copy of the Records of the Services of the 27th Regiment be transmitted to the Horse Guards without further delay, in obedience to your circular letter of the 6th November, 1822; and in reply I have the honour to state to you, for his Lordship's information, that on my succeeding to the command in 1825, the Record Book of the regiment was blank, and not one document in its possession to assist me in furnishing an account of its services, but that when lately on leave of absence, I occupied myself in collecting from documents in possession of the family of the late Lord Clarina, as also from some of the old officers of the corps, a good deal of information of its early history, which I have compiled since my return to the service companies, a copy of which shall be transmitted by the earliest opportunity.

"I have the honour to be, Sir,

"Your very obedient humble servant,

"J. HARE,

"Lieut.-Colonel commanding 27th Regiment."

In accordance with his promise, accompanying his letter dated from St. Vincent on 27th January, 1829, Lieutenant-Colonel Hare forwarded a copy of the Records obtained from the family of the late Lord Clarina and the old officers of the regiment, which he thought would "be found to be correct." It was as follows:—

"*Extract from the 'Memoirs of Captain Carleton.'*

"In May, 1692, Colonel Tiffin's regiment (27th) lay in garrison at Portsmouth, and about two months after was shipped off with other regiments on a land expedition, under the command of the Duke of Leinster, which was intended to proceed against Dunkirk. The French being, however, apprized of the object, the force was directed to land at Ostend.

"The French army, under the Duke of Luxemburg, lay strongly encamped at Steinkirk, near Enghien. King

William, though aware of the many obstacles which lay between the two armies, resolved to attack him (1st August, 1692). Our troops were forced to hew out their passage for the horse, and several pieces of cannon were taken from the enemy; but the further we advanced the ground was found to be such as not to admit of our army being drawn up in battalions, and the troops behind were of course unable to give timely succour to those engaged, and the cannon we had taken were unavoidably left behind.

"Many brave and excellent officers were lost in this battle, among whom were General Mackay (of the — Regiment), Sir John Lainer (of the 1st Dragoon Guards), the Earl of Angus (of the 26th Regiment, with both of his field-officers), Sir Robert Douglas (of the Royal Regiment of Foot), Colonel Hodges (of the 16th Regiment of Foot.)

"I remember one particular action of Sir Robert Douglas, of the Royal Regiment of Foot, who on seeing his colours on the other side of the hedge, in the hands of the enemy. leaped over, slew the officer who had them, and then threw them back to his regiment, redeeming his colours at the expense of his life. Thus the Scotch commander improved upon the Roman general; for the brave Posthumius cast his standard in the middle of the enemy for his soldiers to retrieve, but Douglas retrieved his from the middle of the enemy without any assistance, and cast it back to his soldiers to retain, after he had bravely rescued it out of the hands of the enemy.

"From hence our regiment (27th) received orders to march to Dixmuyd, where we lay some time in fortifying the place. Having brought the fortifications into some tolerable order, we received a command to re-embark for England; and upon our landing directions met us to march for Ipswich, where we had our quarters during the winter of the year 1792. From thence we were ordered to London, to do duty in the Tower. While on this duty the battle of Landen took place, on the 27th July, 1693, and we were soon removed, to our satisfaction, from the *pacific* station to one more active in Flanders.

"Notwithstanding that fatal battle, the confederate army

under King William lay encamped, in the year 1694, at Mont St. André, an open space and much exposed, while the French were entrenched at Uignamont, a little distance from us. The French kept close within their entrenchments, and King William, finding he could not draw them to a battle, suddenly decamped and marched directly to Pont Espiers, by long marches, with a design to pass the French lines at that place. But, notwithstanding our army marched in a direct line, we found that the enemy, sensible of the importance of the post, and of the necessity of securing it, had by the most extraordinary exertions, and with a great loss in men and horses, obtained possession of it, thus preventing the confederate army from entering into French Flanders.

"King William, thus disappointed in his noble design, gave immediate orders for his whole army to march through Oudenarde, and then encamped at Rosendale. After some little stay we were removed to the Cameslins, between Newport and Ostend, and once more took our winter quarters there.

SIEGE OF NAMUR.

"In 1695 the strong fortress of Namur, which was taken by the French in 1692, and since made by them much stronger, was invested by General the Earl of Athlone. After many vigorous attacks, with the loss of many men, the town was taken, the garrison retiring into the castle, into which, soon after, notwithstanding all the circumspection of the besiegers, Marshal Boufflers found means, with some dragoons, to throw himself.

"While King William was thus engaged in that glorious and important siege, Prince Vaudemont being posted at Watergaem, with about 50 battalions and as many squadrons, Marshal Villeroy laid a design to attack him with the whole French army. The prince prepared accordingly, and gave us orders to fortify our camp. The army under the Prince Vaudemont scarcely amounted to 30,000, and Marshal Villeroy was known to value himself upon having 100,000 effective men. The prince sent away all our baggage to

Ghent, but still made show as if he resolved to defend himself to the last extremity in our little entrenchments. The enemy upon their side began to surround us, and in their motions for that purpose blew up little bags of gunpowder, to give the readier notice how far they had accomplished it. There was a large wood on the right of our army, through which lay the road to Ghent, not broader than to admit of more than four to march abreast. Down this road the Prince Vaudemont continued to move off his forces, and he acquired more glory by this retreat than an entire victory might have procured for him.

"With a view of obliging King William to raise the siege from before Namur, Villeroy entered into the resolution of bombarding Brussels, in order to which he encamped at Anderlecht, and there made his approaches as near as was convenient to the town. There he caused to be planted thirty mortars, and raised a battery of ten guns to shoot hot bullets into the place. For five days the bombardment continued, and with such fury that the centre of that noble city was laid in rubbish. After having almost destroyed that fine city, Villeroy, finding he could not raise the siege of Namur by that vigorous attack upon Brussels, decamped from before Brussels, and put his army on the march towards Namur, to try if he would have better success by exposing to show his pageant of 100,000 men.

"The Castle of Namur had been all this time under the fire of the besiegers' cannon, and soon after the army under Prince Vaudemont had arrived, a breach being made in a new work, called Terra Nova, which had been added by the French to the fortifications since 1692, it was resolved in a council of war to attempt a storm. Four entire regiments, in conjunction with drafts from several other corps, including a party from Colonel Tiffin's (27th Regiment), of which I had the command, were ordered for this duty. We were all to rendezvous at Salsines, under the command of Lord Cutts; the signal when the attack was to be made being agreed to be the blowing up of a bag of gunpowder upon the bridge of boats that lay over the Sambre.

"As soon as the signal was made we marched up to the

breach, receiving as we advanced the full fire of the Cohosse Fort. But as soon as we came near enough to mount, we found it vastly steep and rugged. Notwithstanding all which, several did get up, and entered the breach, but not being supported as they ought to have been, they were all made prisoners. This, together with a wound which Lord Cutts received, after we had done all that was possible, necessitated us to retire with the loss of many of our men. Marshal Villeroy lay all this time in sight without making attempt to incommode the besiegers, and at length withdrew his army on the morning of the 20th August, 1695, towards Pemys.

"The retreat of the French army leaving the garrison of Namur without any hope of relief, those in the castle immediately capitulated. After one of the gates had been given up, and Count Guiscard was marching out at the head of the garrison, and Marshal Boufflers at the head of the dragoons, the latter was arrested by order of King William, in reprise of the garrisons of Dixmuyd and Deynse, who, contrary to the cartel, had been detained prisoners, and he continued in arrest until the prisoners made at those places were restored.

"King William, after the allied forces had taken their winter quarters, left Loo for the Hague, and returned to England about the middle of October, 1695.

"**1696.**—About the beginning of 1696 a plot was discovered for the assassination of King William. Soon after the discovery had been made by persons actually engaged in that inhuman design, the 27th Regiment, with several others then in Flanders, received orders to embark for England with all expedition. On our arrival at Gravesend fresh orders met us to remain on board the transports till we had further directions. So soon as the depth of this plot was fathomed, and the intended evil provided against, as well as prevented, King William went over to Flanders in May, 1696, and our regiment received orders for its immediate return. We had little to do except marching and counter-marching, until it was resolved, in a council of war for the better preserving of Brussels from similar attacks to those it had sustained from the French during the siege of Namur, to fortify Anderlecht, upon which duty our regiment as well as others were ordered, and the whole army was under movements to cover that resolution.

"**1697.**—The Peace of Ryswick, which was concluded on the 20th September, 1697, terminated hostilities between England, Spain, Holland, and France; and on 30th October following, between France and the Emperor of Germany. The 27th Regiment (then Tiffin's) was ordered to England, and shortly after to Ireland."

BRISTOL.

We must now pass on to the year 1736. The 27th was stationed at Bristol. A very serious riot happened that year amongst the colliers at Kingsdown. The regiment was marched against them, and several lives were lost. The cool forbearing and excellent conduct of the soldiers on that delicate service received the highest applause, and the public thanks of the magistrates and gentlemen were conveyed to the regiment through the general commanding.

WEST INDIES.

In the year 1739 Spain complained against the maintenance of an English squadron in Mediterranean waters, and insisted on their right to search all British vessels in American waters. The British nation were in a mood for war, and war was proclaimed on the 19th of October.

1739.—The 27th, then Lord Loudon's regiment, went out with Lord Wentworth's expedition to the West Indies, and landed at Bastementos, near Portabello, on the Isthmus of Darien. The climate proved so destructive to the troops, that nothing effectual could be done, and after contending two years against sickness and death, the regiment returned to Europe with nine privates, who alone out of six hundred had escaped the ravages of the fever. In December, 1740, the regiment returned to England.

FALKIRK AND CULLODEN.

We next find them in Scotland, engaged in the war against the Pretender.

1745.—Before the regiment was again completed the rebellion of 1745 broke out in Scotland, and the 27th joined the British forces under Lieutenant-General Hawley. In the memorable action at Falkirk the regiment sustained great loss.

1746.—Soon after the battle of Falkirk, his Royal Highness the Duke of Cumberland took command of the king's forces in the north, and pursued the flying rebels as far as Culloden House, where the Pretender's army was formed to receive him. In the action of Culloden the 27th bore also a distinguished share.

The gallant conduct of Lieutenant Massey, afterwards Lord Clarina, at the head of the grenadier company, attracted his Royal Highness's attention, who in testimony of approbation promoted Lieutenant Massey in the field to a company in the regiment.

A remarkable circumstance occurred in the regiment at the battle of Falkirk. Sir Eyre Coote, afterwards commander-in-chief in India, was then an ensign in the 27th, and on the retreat of the royal army he escaped with the king's colour, which he carried in the action, to Edinburgh. For this he was tried by a court-martial and "broke," but it appearing afterwards that his sole motive was to save the colour, he was restored to the service.

AFFAIR AT ALLOWAY.

In October, the 27th, then Blakeney's regiment, attacked the rear of a detachment of rebels who were conveying stores, &c., at Alloa, and took from them a great booty of cows, horses, baggage, arms, and some money.

General Blakeney being made governor of Stirling Castle,

the 27th, under their brave colonel, sustained an obstinate siege in that fortress, and saved that important place from falling into the hands of the rebels.

IRELAND.

The regiment, being on the Irish establishment, was sent to that country towards the end of the year 1746. It is unnecessary to follow the regiment through all its changes of quarters in Ireland till the year 1756, during which period it was stationed in different parts of the country.

"THE CASTLE" ON THE COLOURS.

In July, 1751, by an order of George II., the regiment was allowed to wear a castle with three turrets on their colours, St. George's colours flying in a blue field, with the word "Inniskilling" over it.

I know not whether there had prior to this date been a castle at all or not on the colours, but the castle as described above still remains on the colours and appointments, retaining the style of spelling Enniskillen at the period of the regiment's institution.

REDUCTION OF CANADA.

1756, June.—War being declared against France in May, 1756, a body of troops—the 27th forming a part—were embarked under the command of the Earl of Loudon, who was appointed commander-in-chief of the army in North America, and landed at New York in June, 1756. An active war was now expected, but much valuable time was lost in making preparations and accustoming the troops to what were called the usages of war, and the years 1756 and 1757 were undistinguished by the acquisition of any object and no opportunity of distinguishing themselves was afforded to the troops; and, excepting the abortive expedition designed against Louisburg, no enterprise was undertaken against the enemy, the troops remaining inactive at York and Albany.

Lord Loudon having been recalled, the command of the army devolved upon Lieutenant-General Abercrombie.

TICONDEROGA.

1758.—The campaign of 1758 opened with brighter prospects. Admiral Boscawen was appointed to command a large naval armament, which with a military force of 52,000 men, of whom 22,000 were regulars, afforded the best hopes of a vigorous and successful campaign. Three expeditions were proposed for this year against Louisburg, Ticonderoga, and Fort du Quesne. That against Ticonderoga was under the command of the commander-in-chief in person. The troops* allotted to this enterprise were embarked in boats on Lake George on the 6th July, and landing without opposition, were formed in two parallel columns. In this order they marched, the enemy abandoning their outposts without a shot. The right column falling in with a detachment of the enemy, a smart skirmish ensued, in which Lord Howe, who commanded the reserve, was killed.

Ticonderoga is situated on a point of land between Lake Chamberlain and Lake George, and nearly surrounded with water. The front of that part which is accessible was strongly fortified with high entrenchments, supported and flanked by batteries, and blocked up with felled trees, with their branches turned outwards, forming altogether a most formidable defence. The garrison consisted of 5000 French troops. The picquets commenced the assault, followed by the grenadiers of the army, and supported by the reserve and 1st Brigade. The troops marched up to the formidable entrenchments and defences without artillery. No ladders were even provided for scaling the high breastworks, and the

* 27th, 44th, 46th, 55th, 1st and 4th battalions 60th, Lord John Murray's Highlanders, with 9000 Provincials and a large train of artillery.

soldiers were obliged to climb up on each other's shoulders, and by fixing their feet in holes made with their bayonets in the face of the work, until the defenders were so well prepared, that the instant a man reached the top he was thrown down. Unexpected and disheartening as these obstructions were, the troops displayed the greatest resolution, though exposed to a most destructive fire from an enemy well covered and enabled to take deliberate aim with little danger to themselves. After persevering for hours under such disadvantageous and disheartening circumstances, the general, despairing of success, gave orders for a retreat, which they effected in good order. Unfortunate as the result of that affair was, yet the nation was highly satisfied with the conduct of the army.

CANADA.

The reader should not confound this war against France with the war with the American States. England's destinies at the period of the outbreak were guided by the Duke of Newcastle. Everything was mismanaged, and the king, George II., who paid more attention to his petty duchy of Hanover than his kingdom, was compelled to have recourse to the immortal Pitt, under whose genius the glory of England rose to great eminence. The famous Seven Years' War had commenced. In fact, England had war in every quarter of the globe. Wolfe was sent out to America by Pitt, and his foresight was rewarded, for Cape Breton and St. John's Island were wrested from the French. They also lost Gorse in Africa. At this time Clive was laying the foundation of our Indian empire.

1759.—The campaign of this year comprehended three very important enterprises. Major-General Wolfe was to attack Quebec from Lower Canada, while General Amherst (now the commander-in-chief, and successor of General Abercrombie) should endeavour to form a communication and co-operate with him through Upper Canada, and General Prideaux to proceed against Niagara.

The army (consisting of 1st Royals, 17th, 27th, two battalions 42nd, two battalions 55th, 77th, with nine battalions of Provincials, in all 14,500 men) under the commander-in-chief was first put in motion, and encamped on the 21st June on Lake George, on the spot where General Abercrombie had encamped the preceding year, previously to the attack of Ticonderoga. The enemy at first seemed determined to defend this formidable post, which had already proved so disastrous to our troops. On the advance, however, of the British army, they abandoned it after a slight resistance, and retreated to Crown Point.

The regiment continued under General Amherst's immediate orders in his subsequent operations against Crown Point, Isle aux Noix, Niagara, at the taking of Montreal, and at the reduction of Canada. Major Massey, who had before been rewarded with a company by his Royal Highness the Duke of Cumberland, for his distinguished conduct at the field of Culloden, fully confirmed by his conduct in these engagements the high opinion which his Royal Highness entertained for his military talents.

At Ticonderoga his conduct was so conspicuous as to obtain the public thanks of the commander-in-chief, who, in addition to this testimony of his approbation, promoted Major Massey to the lieutenant-colonelcy of the 46th Regiment, and shortly afterwards removed him back to his own corps, knowing his great partiality for the Inniskillings.

1761.—The province of Nova Scotia had become disaffected to the mother country, and Colonel Massey was sent over with the 27th Regiment to take the chief command there. This service of the regiment was an arduous one, and so well performed, that the enemy was completely driven out of the interior, and defences established by Colonel Massey and the 27th which remain to the present day.

The object for which so great a force had been sent to North America being accomplished by the reduction of Canada, a large portion of the troops were ordered to the West Indies. Eleven regiments having embarked at North America, arrived at Barbadoes in December, under the command of Major-General Monckton.

1762, January.—This armament sailed from Barbadoes on the 5th; and 8th January, 1762, the fleet anchored in St. Anne's Bay, Martinique, and the troops effected an immediate landing with little opposition.

February.—On the submission of Martinique, on the 16th, to the British arms, the 27th was embarked under Brigadier-General Walsh for the capture of Grenada, which island was also taken possession of without opposition.

WAR WITH SPAIN.

May.—Great Britain having declared war against Spain, it was determined to attack the Havannah, and the command of this important enterprise was entrusted to Lieutenant-General the Earl of Albemarle and Admiral Sir George Peacock. The fleet destined for this service, which now included nineteen sail of the line, eighteen frigates, and some smaller vessels of war, with the troops on board, sailed from Martinique on the 17th of May, and on the 5th of June arrived in sight of Havannah.

June.—The army landed on the 7th, and after a variety of operations on both sides, the British army were put in possession of the Havannah on the 14th of August, nine days after their landing in Cuba. Brigadier-General Walsh's brigade was one of those employed against the Moro Fort, and its loss was considerable, not so much by the enemy as from the effects of the climate and the arduous duty of the siege. After the important conquest of the Havannah, immediate preparations were made for embarking the disposable troops. The 27th, being greatly reduced in numbers, was one of the regiments ordered to New York, where it landed in October.

"Head-Quarters, New York, October, 1763.—General Orders.—The commander-in-chief was extremely glad to hear that the Enniskillen regiment, quartered at Three Rivers, had behaved so like good soldiers, upon the order for the stoppage of 4*d.* for provisions. Their submission to that order was becoming of soldiers, and the silence on that occasion was meritorious." In September, 1765, the regiment was distributed as follows:—Four companies in the town of Quebec, three companies to Trois Rivières, and two to Montreal. In March, 1766, the regiment was ordered to Montreal, there to be quartered, Colonel Massey being appointed to command the district.

In August, 1767, the regiment embarked on board the transports for Europe, September 29th landed at Cork, and next day proceeded to Dublin. In the following year the regiment was quartered at Limerick, returning to Dublin in 1769, where it remained until 1774, when it returned again to Limerick, and in August, 1775, was ordered to prepare for foreign service, the recruiting parties and absentees being called in.

THE AMERICAN WAR.

George III. was now king. The Hon. George Grenville had become Prime Minister, and to meet the expenses of the late war he had imposed taxes on certain papers and parchments used in America. The American colonists refused to pay taxes, since they had no members in the British Parliament. Grenville resigned, and the Marquis of Rockingham, his successor, repealed the Stamp Act. The Duke of Grafton, and Pitt, Earl of Chatham, next came to office, and in spite of the warnings of Pitt new taxes were imposed on the colonies. In 1768 Chatham resigned, and in 1770 gave way to Lord North, under whose ministration the taxes were kept up, and by whom England lost the United States. The States resisted the taxes; Chatham and Burke warned the English Commons in vain, taxed tea was thrown into Boston harbour, and at last, after ten years of word-warring, real war commenced. The battle of Bunker's Hill taught the English soldiers that their brothers in America were of equally stern metal.

The 27th sailed from Cork on the 26th of September, and arrived at Boston in the end of October. In March, 1776, they proceeded to Halifax, on the evacuation of Boston by the British army. In July, landed with the army from Halifax at Staten Island, where the forces destined for General Howe's command were united. August.—The campaign opened by a landing on Long Island. In the action near Brooklyn, the 27th was on the right wing of the army, in the division of General Clinton, which was principally engaged. The regiment was also engaged in the subsequent operation of the army against New York, in the battle of White Plains, in the affairs of Fort Washington and King's Bridge, and at the reduction of Rhode Island.

1777.—During this year the 27th was not engaged in any active operations, except in some foraging expeditions, it being with the division of the army at Rhode Island under Major-General Sir Robert Pigot, and which was not employed in the campaign of this year.

1778, March.—Employed under Colonel Mawhood in an expedition to the Jersey coast, in the affairs of Quintin's Bridge and Hancock's Bridge. July.—On the evacuation of Philadelphia returned with General Clinton to New York. November.—Count D'Estaing, with the French fleet, having quitted Boston and sailed for the West Indies, Major-General Grant with ten regiments, the 27th one of them, were despatched by Sir Henry Clinton for the protection of the West India islands. They sailed from Sandy Hook on the 3rd of November, under Commodore Hotham. December 10th, arrived safe at Barbadoes.

ST. LUCIA.

The place of debarkation for the capture of St. Lucia being with Admiral Barrington, the fleet sailed from Barbadoes on the 10th of December. On the evening of the 13th, the army under Major-General Grant landed at the Grand Cul de Sac Bay, and on the morning of the 14th took possession of Mount Fortune with all the stores and magazines belonging to the island. On the 15th Count D'Estaing, with the French fleet, stood in for the Carenage, and on the 18th landed with 9000 men collected from the French islands. On the 19th he attacked General Meadow's Brigade, the reserve consisting of the 5th Regiment and flank companies of the army, strongly posted on the Vigie, with 5000 French troops in three columns, who were repulsed with great loss, and retreated in confusion. Monsieur D'Estaing having given up the contest, embarked his troops on the 28th of the same month, and sailed away. Monsieur de Micond and the inhabitants immediately capitulated to the British army.

1779, July.—The regiment embarked at St. Lucia for the relief of Grenada, and was present at the sea action of the 6th July off Grenada, between the French and British fleets under Count D'Estaing and Admiral Byron. Having failed in the attempt to succour Grenada, the troops sailed for St. Christopher's. Shortly afterwards the 27th returned to St. Lucia.

1780.—In garrison at St. Lucia. The regiment in this and the preceding year suffered severely from fever, the loss of officers and men being very great.

1781.—Embarked at St. Lucia under General Faughan for the capture of the Dutch island of St. Eustatius, but the employment of so large a force as was embarked not being required against that place, the 27th and some other corps were ordered to Barbadoes.

1782—1784.—The regiment continued in the West Indies, which at that period was the most active theatre of war for the British fleet, and various enterprises were undertaken with various successes. On the peace of 1783 the establishment of the regiment was reduced to eleven companies, ten for foreign service and one for recruiting.

This was a time of great men, great achievements, and great disasters. The great Chatham was sinking, but Richard Brinsley Sheridan, Edmund Burke, young Pitt, and Henry Grattan were rising in fame. Rodney had captured Spanish convoys, and added laurels to England's fame by his victory off Cape St. Vincent and defeating Count de Grasse. Lord Cornwallis had been compelled to surrender to the Americans at Yorktown with 7000 men. The French retook St. Eustatius. We lost Demerara and Essequibo, with St. Kitt's, Nevis, and Montserrat; and, to crown all, the Island of Minorca was surrendered. England at this time was, like the rest of Europe, exhausted, and on January 20th, 1783, the Treaty of Versailles gave rest. The independence of America was acknowledged; Britain retained Gibraltar from Spain, which received Minorca and both the Floridas. England restored St. Lucia and ceded Tobago to France, receiving in return Grenada, St. Vincent, Dominica, Nevis, and Montserrat. I mention only these portions of the treaty, as they are principally connected with our subject.

1785.—Continued at Barbadoes. In March returned home from the West Indies.

1786–1788.—It is unnecessary to follow the regiment through its many changes in those years of quarters, which were chiefly in Scotland and Ireland.

IRELAND.

1789–1791.—The 27th was quartered in Dublin, being retained there an extra year for the purpose of instructing the new garrison in the Dundass system of field exercise and movements, which were then ordered to be established in the army, and in which the 27th had been perfected by General Dundass himself.

1792.—And it is worthy of a place in this Record of the regiment to note that such was then its high state of order and perfect discipline, that no less than fourteen of its sergeants, from 1790 to 1793, were promoted to adjutants and sergeants-major of other regiments of the line and militia.

1793.—Removed from Dublin to Galway, thence to Cork. Hostilities having been declared against France, the regiment was ordered to embark at Cork, to join the Duke of York in Flanders.

WAR WITH FRANCE.

By this last sentence it will be seen that the peace of Versailles was not very lasting. The conduct of the French Assembly, wresting Avignon and the Venaissin from the Pope, the Revolution with its horrors and the promising armed assistance to the disaffected of all nations, left France without an ally. Europe was shaken to its centre. The British Government had no alternative but war. They did not accomplish much.

August.—Landed at Ostende, and joined the army of his Royal Highness, then encamped in the neighbourhood of Menin.

This camp was soon broken up, and his Royal Highness proceeding with the combined army to join the Prince of Saxe Coburg, then before Maubeuge, four British regiments were ordered back to England (the 27th one of them), to join an expedition then preparing under Lieutenant-General Sir Charles Grey against the French colonies in the West Indies. While those regiments lay on board ship in the harbour of Ostende, the enemy appeared before Nieuport, and pressed that town, which was garrisoned by one British regiment (the 53rd) and a Hessian battalion, so vigorously, that it was found necessary to disembark the troops at Ostende for the relief of that place. On the appearance of this reinforcement the enemy retired with great expedition, leaving several pieces of cannon, mortars, and ammunition. After this performance the troops re-embarked for England. On arriving at Portsmouth, the destination of the regiment was changed from an expedition to the West Indies to another forming against the coast of France, under the command of Lord Moira.

The expedition under Lord Moira sailed from Portsmouth on the 20th of November, and the next day reached the coast of France. Having cruised to the eastward of Cape La Hogue, the transports were, in consequence of bad weather, obliged to put into Guernsey, where they remained until the beginning of January following, when the whole of the troops returned to Portsmouth. The 27th Regiment marched to Southampton in Lord Cathcart's Brigade, and was there completed in its establishment by drafts from the 103rd Regiment.

1794.—The regiment continued in Southampton till the beginning of June, when an encampment was formed on Netley Common, in Hampshire, under the command of Lord Moira.

June.—The original destination of this force being changed, in consequence of its being found necessary to reinforce the Duke of York, this camp broke up on the 18th of June, and the troops again embarked to join his Royal Highness, and landed on the 26th of the same month at Ostende, numbering 7000 men. On the evening of the 28th the troops moved from their position on the Sand Hills, in the neighbourhood of Ostende, and, after a great deal of forced and fatiguing marching, caused General Vaudamme to retire from his position near Bruges with very superior numbers. Joined the Duke of York's army on the 9th of July at Malines. The march of Lord Moira's division from Ostende was a most fatiguing one, but it had been so well conducted that the enemy, although in very superior numbers, under General Vaudamme, did not venture upon any attack, or attempt to impede the march of the British troops, except by a trifling dash of his cavalry into Alost.

FLANDERS.

October.—A variety of movements and some smart skirmishing, but nothing of any importance took place till the beginning of October, when the army crossed the Waal, at Nijmegen, and took up its position on that river. Several smart but partial actions took place in that position, till the morning of the 20th, when the enemy made a general attack on all the advanced posts of the army, which were defended with great gallantry, and the enemy repulsed; but the outposts being at last driven in by a system of constant attacks with very superior numbers, the enemy established themselves in front of Nijmegen, and commenced to erect batteries against the place. The 27th was one of the corps ordered to garrison it. On November the 4th, it being resolved to attempt the destruction of the enemy's works by a sortie from this garrison, General De Burgh, with six British regiments of infantry, two of dragoons, two battalions of Swiss in the Dutch service, and some Hanoverian horse, were ordered on this duty. The 27th was posted on the right of the British line; in that attack the enemy made a brave defence, but the works were carried with great gallantry. On the 7th it was found necessary to evacuate the town, as the enemy continued their approaches with fresh vigour, and were reinforced by the divisions called the Army of the North. A picquet of grenadiers of the 27th Regiment, commanded by Captain Warren, was the last of the British that left the town. This picquet was stationed about half a mile in advance of the Belvidere gate, and was the only force outside the works of the town. After destroying four guns in the battery in front of the gate, and having brought away the keys of the garrison, this picquet with great difficulty crossed the bridge about three o'clock in the morning, but not without the loss of some men, who were drowned in endeavouring to cross the river on the flying bridge.

After the evacuation of Nijmegen the army was cantoned along the Waal, where they suffered much from the severity of the weather. So intense was the frost, that the enemy were enabled to cross the Waal on the ice, and, by availing themselves of their superior numbers, to commence active operations. On the 28th of December, the river being completely frozen, they crossed in considerable force, under Generals Dundass and Osten, near Brummel. The whole of the division of Dundass, about 8000 British, the 6th or Lord Cathcart's Brigade forming a part, advanced against them. The French took up the most favourable position at Thuyl, flanked by batteries planted in the Island of Bammel; but notwithstanding these obstacles, and their great superiority of numbers, they were driven from all their posts, and obliged to recross the Waal.

1795.—On the 8th of January they again crossed in the

immediate vicinity of the 6th Brigade, which was instantly in action. This brigade had a busy day, and General Walmsden's orders and despatches mentioned that the British regiments under Lord Cathcart behaved with extraordinary gallantry. The attacks were made and received with such energy, that each party alternately attacked and was repulsed four times successfully, till at length the enemy was forced to give up and retreat with considerable loss. The 27th on this day took from the enemy an 8-pounder. The army crossed the river Leck, in front of Geldermelsen, in face of the enemy's whole force, and under a dreadful fire of round and grape shot, which killed and wounded eight officers, and seventy non-commissioned officers and privates. Among the killed was Lieutenant-Colonel Buller. On the 10th the enemy again crossed with his whole force in five columns; and on the 14th, Pichegru made a general attack along the whole line, from Arnheim to Amerongen, when the British troops, after a resistance which continued till night, withdrew at all points, and commenced a retreat.

In the succeeding disastrous and harassing retreat of the British army to Deventer, the 6th Brigade of British infantry, a brigade of British cavalry, and some foreign corps, and a large field train of artillery, formed the rear-guard of the right wing of the British army. Frequent skirmishes with the enemy took place during the retreat, and the rear-guard was so closely pressed, that in the beginning of March the greater part of its baggage, with all its women and sick, fell into the hands of the enemy.

RETREAT THROUGH HOLLAND.

The following is an extract from "A Concise Narrative of the Retreat through Holland to Westphalia, in the years 1794 and 1795:"—"It was now determined by a council of war to abandon the position on the Waal entirely. Accordingly, several heavy guns having been spiked, and great quantities of ammunition destroyed on the 4th, the troops on the 6th fell back upon the Leck, part of the army crossing the Rhine at Rhenen. A sudden thaw unexpectedly affording some prospect of preserving the posts on the Waal, orders were immediately issued to the troops who had not passed the rivers to remain in their cantonments, and for the others to countermarch. Lieutenant-General Abercrombie's and Major-General Hammerstein's corps, with some Austrian battalions, were to have pushed forwards towards Bommel on the 7th; and, to co-operate with them, General Dundass was directed to occupy Buren and the heights near it, early on the morning of the 8th. The 14th and 27th Regiments were immediately detached from Culenberg to retake Tiel; but on their arrival near Buren, they found the enemy advancing upon them in force; and Lieutenant-Colonel Buller immediately took possession of the town, waiting till the arrival of the head of General Dundass's column, when Lord Cathcart, having previously reconnoitred, found the detachment of the enemy at Geldermalsen did not exceed 800 men, with some hussars, and one piece of artillery, and he consequently immediately determined to dislodge them. This was effected in a very spirited manner by the 14th and 27th Regiments; driving in the advanced posts at Buremulsen, they pursued to Geldermalsen,* and carried that village at the point of the bayonet, seizing upon the French gun, a long 8-pounder. It was, however, absolutely necessary for them to make as expeditious a retreat as possible, which was accomplished in a steady and soldier-like manner, covered by the 28th Regiment."

* A small detachment of the British Hulans were the only cavalry engaged on the 8th. They pursued the French to Buremulsen, charged across the Ningen, *on the ice*, with the most daring intrepidity, and brought off several prisoners. The British and French *at Geldermalsen* repulsed each other *four times* in the course of the day, and the gun that was taken by the 27th was sunk in the river, the ice breaking under it. Lieutenant-Colonel Buller, of the 27th, was mortally wounded. Lieutenants Conner, Norbory, and Ensign Kelly, of the same regiment, were killed upon the spot. Lieutenant-Colonels Gilman of the 27th, and Hope of the 14th, were wounded, the latter very severely (but he has since recovered); also Brigade-Major Wilson of the 27th, Captain Perry of the 14th, and Lieutenant Raitt of the 42nd. Eleven rank and file and three horses were killed; three sergeants and 111 rank and file wounded, and seven missing. Captain Perry's wound proved mortal.

April.—The regiment embarked and sailed for England, and on its arrival at Portsmouth was quartered in Dorchester Castle.

In July, 1795, the 27th embarked at Portsmouth with four other regiments, under General Graham Campbell, for the purpose of co-operating with a body of emigrants against Chouans, which had made a descent at Quiberon, under M. de Puisage and Count Sombreuil. The ships were long beating down the Channel, and having at last reached the French coast, they fell in with the *Anson* frigate, which acquainted them of the disastrous event at Quiberon. The troops in consequence returned to England; the 27th to Southampton, where it encamped. The regiment was there again completed to their establishment by drafts from several of the newly raised corps.

Sir Ralph Abercrombie having assumed the command of a numerous armament then preparing for an expedition to the West Indies, the 27th was again embarked for that service. November 11.—The embarkation of the whole of the troops being now complete, the fleet, amounting to upwards of 300 sail, got under weigh with a favourable breeze; but in consequence of the flag-ship *Impregnable* having struck on a sand rock, a signal for recall was made, and the fleet was ordered to come to anchor. On the 15th November, the fleet again weighed anchor and sailed; but scarcely had it cleared the Channel, when it was dispersed and driven back by a furious gale, and with the loss of several ships and many hundred lives. Another attempt was made to put to, and as the fleet was clearing the Channel on the 13th a violent storm commenced, and continued with unabated violence for many weeks. The admiral (Christian), however, persevered until the end of January, when the disabled state of such of the ships as kept with him rendered it impossible to remain longer at sea; he therefore made a signal to run for Portsmouth, where he arrived on the 29th January, 1796. The transports of the 27th anchored at Portsmouth the same day, after being upwards of seven weeks at sea.

ST. LUCIA.

On the 4th of March the fleet under Admiral Christian finally sailed and arrived at Barbadoes, and next day appeared off St. Lucia.

The 27th Regiment landed on the 9th of May, 1756, and it was, on the 21st of May, placed under the command of that great officer and superb soldier, Sir John Moore, a major-general in Sir Ralph Abercombie's army.

Major-General Moore was employed by Sir Ralph to conduct the siege on the side of Camp de Chasseau à Ferrauds. The investment, from the hilly and bushy nature of the country, was difficult, and the post unconnected. Moore's post was the left of the right wing, and the most important of the whole, being, in fact, the only one from whence the place could be effectually approached. A great many partial but sharp affairs happened in the course of this siege, not all to the advantage of the British troops, who were generally ill-disciplined, and in several instances very ill-commanded; wherefore the actions of the 27th now to be noticed stand out in greater relief; and the facts being taken from the journal of their general, wherein he has not failed to censure severely those who failed, speak for themselves.

On the 24th of May, at daylight, all the battalions opened against Morne Fortuné; but about six o'clock some of the guns were turned on the enemy's most advanced post, and especially on a *flêche* to the right of it, to disperse any troops hidden in the thick brushwood about. Then the grenadiers and light company of the 27th, under Lieutenant-Colonel Drummond, advanced, and General Moore placed himself between the two companies, which were, on reaching the top of the hill, fired upon by thirty or forty men in the *flêche.* No resistance had been expected, and Sir Ralph's orders were not to assail the *flêche* itself, which was exposed to the shells and grape from Morne Fortuné, but to effect a lodgment on the reverse side of the hill, until a covered way of communication could be made. Moore, however, saw that it was requisite to take the *flêche;* and the two companies stormed it, with the loss of only two men.

The summit was open, the sides covered with brushwood, and the general proceeded to establish his lodgment under discharges of grape from the enemy; but the working parties previously ordered did not come up, and, the enemy's fire having ceased, Moore went to the rear to enforce this necessary co-operation.

There he met Sir Ralph, and while speaking to him, troops were observed from Morne Fortuné to attack the post just gained. Some guns were turned on them, and Moore, running to the *flêche,* reached it just as the attack commenced. The enemy's fire was brisk, and they took skilful advantage of the ground in skirmishing. The regiment fell fast, and the general directed the flank companies to charge. Colonel Drummond headed them, and cutting down an officer with his own hand, drove back the French with the bayonet, but suffered from grape in returning. The enemy being then reinforced, attacked a second time, when Moore sent a battalion company to line a hedge to his left, where it could secure a dip by which the assailants could come within twenty yards under cover; the company failed to line the hedge, and the enemy took advantage of it, and attacked with great boldness at a point where the only front that could be presented was narrow.

Many officers and men were down, many more falling. The rest displayed great spirit; for their adversaries' fire was superior, and the ground was so bushy the whole of the men could not get into action or support each other. Moore, seeing the soldiers fall so fast, feared they might give way, and knowing Sir Ralph was looking on at no great distance, begged of General Hope, who happened to be present, to go back and entreat Sir Ralph to order the regiment of Royal Strangers—a French corps in our service—to show itself on his left. Hope went, and the combat became more serious. The flank companies had suffered too much to charge again, but Moore, turning to Colonel Gilman, told him if two companies did not charge, the post was lost. Two companies did move out with great gallantry, but could only advance in file to form under fire; and Captain Dunlop was wounded, and fell at their head. Major Wilson was killed immediately afterwards, whilst heading them to the charge; but the enemy, though they had the mastery in firing, could not stand the bayonet, and fled. Moore seized the occasion to make Gilman in person bring up two companies from the rear, where they had not suffered, to line the hedge on the left before spoken of; and he did it well, sending back the companies before employed ineffectually in the service. The enemy then retreated, the action terminated, and Morne Fortuné surrendered.

The commander of the forces ordered that the 27th Regiment, in consequence of their gallant conduct, should take possession of Morne Fortuné, and that the French garrison (about 2000) should march out and lay down their arms to the regiment on the glacis. Sir Ralph further directed that the king's colours of the Inniskilling regiment should be displayed on the flag-staff of the fort for one hour previous to hoisting the usual union flag. After the capture of St. Lucia the regiment was sent to reinforce the garrison of Grenada.

In the first pages of the Records, but not in order of sequence, I found the following:—

"*Abstract of General Orders issued by Lieutenant-General Ralph Abercrombie, commanding the forces in the island of St. Lucia, 26th May*, 1796.

"Parole—Enniskillen. Countersign—Gilman. The 27th Regiment, under the command of Brigadier-General Moore, will this day at twelve o'clock take possession of Fort Charlotte, the present garrison having first marched out and laid down their arms on the glacis to that regiment. Brigadier-General Moore will then plant the colours of the 27th Regiment on the fort.

"The Commander-in-Chief is proud to say that the services which have been performed by Brigadier-General Moore have been auspicious—that it is unnecessary for him to detail them. His conduct in particular on the 24th of May could not but attract the attention of the whole

E

army; and the behaviour of the Enniskillen regiment of infantry, who acted on that day with him, was so worthy of praise, that it deserves the Commander-in-Chief's highest approbation."

Regarding the taking of St. Lucia and the right of having the action remembered on its colours, the following correspondence took place:—

"SIR, "13th February, 1835.

"The opportunity which now offers to me, of your ordering new colours for the Enniskillen regiment, is so favourable that I cannot allow it to escape me, of bringing before you the claims of our corps to have 'St. Lucia' added to the other devices on its colours.

"In 1826 or '27 I handed to our late colonel, the Marquis of Hastings, a copy of the enclosed extract from the Records of the Services of the Regiment, and I was then induced to submit our claims also to this honour, from its having been a year or two before conferred upon the 53rd Regiment, for services performed by that gallant corps in the months of April and May, 1796; but his lordship being then on the eve of embarkation for Malta, I believe it escaped his recollection to lay the claim of the 27th before the Commander-in-chief, and I have since waited your return home to renew it.

"The 27th was in brigade with the 53rd under the distinguished Sir John Moore, and the enclosed copy of Sir Ralph Abercrombie's order will, I trust, show that it reaped also its share of honour on the service.

"I have the honour, etc.,

"J. HARE,

"Lieutenant-Colonel Enniskilleners."

This "copy" states what is not mentioned above:—"The French garrison, 2000 strong, *in compliance with their own request*, were permitted to lay down their arms to the Enniskilleners on the glacis;" and the "UNPRECEDENTED HONOUR" of having the flag of the regiment displayed for an hour before the king's colour "was recognized by three simultaneous cheers from the whole army." I find from another record that General Moore said, "*That post was of the utmost importance, and it was saved by the good conduct and regularity of the 27th Regiment,* though they *laboured under every disadvantage.* Sir Ralph thanked me for my exertions, and said he could never sufficiently repeat his obligations to me. I told him *we owed the post to the gallantry of the 27th Regiment.*" It lost 120 men and 80 officers in this engagement.

Having reference to Lieutenant-Colonel Hare's letter, the following came from the Horse Guards a year after :—

"Horse Guards, 6th April, 1836.

"SIR,

"I have the honour to acquaint you, by direction of the General commanding-in-chief, that his Majesty has been graciously pleased to permit the 27th or Enniskillen Regiment of Foot to bear on its colours and appointments, in addition to any other badges or devices which may have heretofore been authorized, the words 'St. Lucia;' in commemoration of the gallant conduct evinced by the 27th Regiment on the 24th May, 1796, in the attack upon the Morne Fortuné, which led to the surrender of the island of St. Lucia on the 26th of that month.

"I have the honour, etc.,

"JOHN MACDONALD,

"Adjutant-General."

1798.—The regiment embarked at Grenada for England. A particular mark of his Majesty's favour was conferred on the regiment on their being ordered home from the West Indies. The practice then usual in the army of drafting men from one regiment was directed not to take place with the soldiers of the Enniskillen regiment, in consequence, as his Majesty's order stated, of their *distinguished conduct* at St. Lucia.

July.—The regiment arrived at Portsmouth and was sent into quarters at Winchester. In the following month the regiment was ordered to Guernsey. Returned to England in January, 1799, and was quartered in Lyndhurst, where a large reinforcement of recruits joined the regiment from Ireland. In July they joined the camp then forming at Barham Downs, for an expedition to Holland. In August, General Coote's brigade embarked at Margate, and sailed the same day.

CALLANSTERG.

Napoleon was rapidly accomplishing his ends. By the Treaty of Campo Formio, signed on the 17th October, 1797, France had acquired Flanders; and as Britain was still at war with France, she prepared, in conjunction with Russia, a great expedition against Holland. The French were alarmed by the news of the preparations, but they could afford no help to Brune, who had 15,000 French and 20,000 Dutch troops.

After a great deal of adverse weather, the fleet anchored at the Helder, the bomb vessels, sloops, and gun-brigs being stationed close along the shore, to secure the beach and cover the landing. The troops began to disembark by daybreak, General Coote's brigade being the first to effect the landing. The boats with the 27th being ordered in advance, the first fire from the enemy was into the boats containing the grenadier company, of which several men were killed and wounded. General Darndels having assembled a body of infantry, cavalry, and artillery at Callansterg, made many efforts to dislodge the right of the British, composed of the Guards and General Coote's brigade, without being able to make any impression. On one occasion, when the enemy was strongly contesting a ridge of sand hills, Sir Ralph Abercrombie rode up to the 27th, and called out to the commanding officer for a charge from his old friends the Inniskillings. The charge was instantly made; the disputed position was carried, and the enemy driven at all points; but the 27th had a severe loss in being deprived for the remainder of the campaign of their brave and distinguished commander, Colonel Samuel Graham, who in the charge received a severe wound in the head, by which he lost an eye.

The conduct of the grenadier company on that occasion was particularly distinguished, and they were thanked by General Coote in the field. The loss in this battle was one officer killed, four sergeants and forty rank and file killed and wounded.

BERGEN.

1799.—The regiment was afterwards engaged in the actions of the 29th September, and 2nd and 6th of October. During the whole of the action of the 2nd of October, General Coote's brigade was kept actively engaged with the enemy near the villages of Schord and Pergen. Towards the close of the action a large body of the enemy having been seen moving to their left, apparently with a view of turning the right of Lord Chatham's brigade, the Queen's, 27th, and 29th were ordered to form on his lordship's right, to support him and extend the line. The 27th was posted at the termination of an avenue leading to Bergen, and was attacked by a large body of the enemy, issuing from the woods. The regiment charged and drove them back into the wood, and no further attempt was that day made by them on the side of Bergen. Three companies which were detached to act on the right of Lord Chatham's brigade engaged and defeated a strong body of French infantry, and were thanked on the field by Lord Chatham.

In the following year, 1800, the 27th was one of the corps encamped at Swinley, when Lord Chatham took an opportunity, on the day the king was reviewing the troops, to point out Lieutenant Standish to his Majesty as an officer whose gallant conduct had attracted his lordship's notice in Holland. His Majesty immediately ordered Lieutenant Standish to be promoted. That officer was afterwards appointed to a company in the 54th. On the evacuation of Holland by the British army, the 27th returned to England, and was quartered in Dover Castle.

FERROL.

In May, 1800, a second battalion was added to the regiment, of volunteers from the Irish militia, who joined at Dover. In June the regiment marched in two battalions from Dover to Swinley Camp on Bagshot Heath; and in August the camp broke up, and the troops proceeded in brigade to Portsmouth for embarkation, sailing on the 6th for Quiberon Bay, where an expedition against Ferrol, under Sir James Pulteney, was assembling.

On the evening of the 25th the troops landed near Ferrol with little opposition, and advanced to the heights which overlook the town and harbour of Ferrol. Next morning a smart skirmish took place; the Spaniards were driven at all points, and retired within the garrison, their gun-boats in the harbour keeping up during the day a heavy and galling fire of round shot. The general commanding ordered the troops to withdraw in the evening, and to re-embark. They were not molested in their retreat. The fleet proceeded next day, the 27th, for Gibraltar, where they joined Sir Ralph Abercrombie early in the month of September.

EGYPT.

Buonaparte had for a long time held that the true com mercial route to India lay through Egypt, and that it was only by its possession and the conversion of the Mediterranean sea into a French lake that he could accomplish his purposes of complete aggrandisement and injury to British power. He set out for Egypt with a magnificent armament in 1798. The battles of the Pyramids and of the Nile followed, and the memorable siege of Acre. The reverses of his armies in Italy and Switzerland caused his personal return in the next year, leaving the command to Kleber. Kleber being murdered in the year 1800 by a fanatic, Menou succeeded to the command of the French army. Britain was now acting in concert with Turkey to expel the French from their usurped settlement in Egypt. It had been arranged that Abercrombie should proceed from Malta, while 8000 troops from Bombay were to arrive at Suez, and the Vizier was to co-operate with a fresh invasion. The Ottoman invasion was useless, and the arrival of Sir David uncertain, so Sir Ralph Abercrombie gallantly determined to make the attempt alone. How he was assisted by the Inniskillings in this endeavour is told hereafter. For its services in this campaign the 27th bears on its appointments a figure of the Sphinx, with the word "Egypt" underneath.

October, 1800.—The fleet with the troops anchored in Cadiz Bay for the capture of that fortress and the Spanish fleet, but that enterprise being abandoned in consequence of a contagious fever which prevailed in that city, the fleet, after watering at Fortuan, sailed for Egypt.

1801.—The battalions of the 27th Regiment having contracted a severe illness in consequence of the length of time embarked (five months), together with the crowded state of their transports, the 1st Battalion was left at Malta to recover, and the 2nd Battalion was ordered back to Lisbon. This battalion recovered so rapidly, that it was again embarked for Egypt, and arrived in time to be present at the landing at Aboukir Bay on the 8th March, and in the advance towards Alexandria on the 13th March. In the battle of the 21st March, the 2nd Battalion, 27th Regiment, was in Lord Cavan's brigade, which was posted on the left of the front line. This brigade was not engaged, as General Reynier, who commanded the left of the French line, never advanced to the attack, but kept up a heavy cannonade, from which all the regiments suffered considerably.

On May 7th, the 1st Battalion, from Malta, joined the army in Egypt. Both battalions were employed before Alexandria, and at the landing at the westward side of Alexandria under Major-General Coote.

On August 22nd, at the advance of Major-General Coote's corps to blockade the westward side of Alexandria, the 1st Battalion 27th formed the advance, with two squadrons of the 23rd Dragoons, and some light field artillery. This advance corps lost some men as it drew near the city. After the reduction of Egypt the two battalions returned to Malta in November, where they were placed in garrison.

France refused to abandon Egypt till forced by defeat. The Treaty of Amiens was signed on October 1, 1801, and both nations were delighted at the prospect of peace. Other treaties were concluded by France with Bavaria,

Austria, and Russia, so a period of universal quietude was hoped for. But though the clash of arms was silent for a time, there were social and political struggles in France. Two years later, Napoleon directed his ambition towards Egypt. War was declared, and a rupture occurred with Spain.

1802.—Immediately after the peace of Amiens the 2nd Battalion was reduced. The establishment of the 1st Battalion was increased in men to 1200, but reduced in officers, the twelve junior lieutenants being placed on half-pay. Two hundred supernumeraries were permitted to be kept on the strength of the regiment upon the consolidation of the two battalions.

1803–4.—In garrison at Malta. A second battalion was again added to the regiment in November, 1804.

1805.—In this year a third battalion was added to the regiment. The 1st Battalion embarked at Malta with the expedition to Naples under Sir James Craig, in co-operation with a body of Russian troops, having the final view of forming a junction with the Austrian army at that time serving under the Archduke Charles in Italy. The occurrences which then took place under that army rendered the object of the British general ineffectual. He therefore embarked his troops, and took possession of Sicily in February, 1806. The 2nd Battalion during this time was serving in Hanover with Lord Cathcart.

FIRST BATTALION.

MAIDA.

July, 1806.—The 1st Battalion was one of the corps employed in the expedition to Calabria under Sir John Stuart. At the battle of Maida, on the 4th July, it was in Brigadier-General Cole's brigade, which was posted on the left of the British line.

At this famous combat, where the British met the French in the proportion of four of the former to six of the latter, the

27th were posted in a field of corn or high stubble, and were shelled by the French, who, with their usual skill, were making good practice, and pitching the shells into the field and amongst the soldiers; the shells bursting set the stubble on fire, and the men had to move out of the reach of the burning stubble.

This movement was seen and mistaken by the general commanding. A staff officer was instantly dispatched, who galloped up to them in hot haste, and called out angrily to Private Scott, who chanced to be the nearest man, and asked with an oath, "Are the 27th about to run?" Scott, indignant at the idea, replied in no soft tones, "Hell to your soul! you or any other ruffian shall never see that day." The officer, well pleased to find the men steady under such circumstances, and at so critical a time, rode back, first remarking to Scott that his answer was exceedingly improper and disrespectful, and that he should hear further about it at some future time. In a month or two after, as the regiment was drawn up on parade, a non-commissioned officer came to Scott, and informed him that he had been made a sergeant. Scott, who could neither read nor write, and consequently never dreamed of promotion, thought his informant was quizzing him. An officer of his company, overhearing their talk, stepped up to them and said, "It is a fact, Scott, you have been made a sergeant, and the reason of promotion is the answer you gave the staff officer at the battle of Maida."

Extract from Sir John Stuart's Despatch.

"The enemy being completely discomfited on their left, began to make a new effort with their right in the hopes of recovering the day. They were resisted most gallantly by the brigade under Brigadier-General Cole. Nothing could shake the firmness of the Grenadiers, under Lieutenant-Colonel O'Callaghan; and of the 27th Regiment, under Lieutenant-Colonel Smyth. The cavalry, successfully repelled from their front, made an effort to turn their left. The manœuvre of the 27th Regiment in throwing back its left wing, and by a heavy and well-directed fire, supported by the 20th Regiment, which Lieutenant-Colonel Ross opportunely threw into a small cover on their flank, entirely disconcerted this attempt. This was the last struggle of the enemy, who now, astonished and dismayed by the intrepidity with which they were assailed, began precipitately to retire, leaving the field covered with carnage."

August.—The army from Calabria returned to Sicily, where a reinforcement of troops had arrived from England, under Lieutenant-General Sir John Moore. The 2nd Battalion, 27th Regiment, was one of the newly-arrived corps.

In 1807 the 2nd Battalion was sent to reinforce the garrison of Malta. In 1808 the 1st Battalion was in garrison in Sicily; the 2nd Battalion at Malta.

1810 and **1811**.—In garrison at Messina and Palermo. In November the 1st and 2nd Battalions embarked at Palermo for the south of Spain; landed in December (in 1812) at Alicante, and were cantoned in that neighbourhood until the following March.

PENINSULAR WAR.

The Records have now brought us to the most interesting military epoch of the present century, the most terrible struggle the world has witnessed since Hannibal strove against Scipio for the mastery of Italian soil, or the distinguished Macedonian contested the sovereignty of Asia at the Granicus and Arbela, and humbled the Indian Porus on the banks of the Hydaspes. Napoleon's genius had laid Europe at his feet, with the exception of Russia, strong in her snowy steppes, and England, secured by her wave-washed shores. Defeating Alvinzi on the plateau of Rivoli, he crushed Italy, and placed his brother-in-law, Murat, on the throne of Naples. The Austrian power was crushed at Wagram and Austerlitz; and Prussia was humbled on the field of Jena. Napoleon had raised his brother Louis to the throne of Holland, and he now sought to make his brother Joseph king of Spain. Unless now checked, there would be no safety in Britain. The Spaniards resisted this act of aggression; so that out of this movement of Buonaparte's sprang the greatest struggle of modern times—the Peninsular War. The maritime power of France had been broken in several engagements by Nelson and his *confrères.* British soldiers were now despatched to Portugal; and on the worshippers of the Emperor and the red-coated lions of England, the eyes of the world were turned, as Wellington thwarted the schemes of Napoleon, and dyed the sunny slopes of Spain with the best blood of Europe. I have referred to the three battalions of the 27th Regiment. The Records first speak of the 1st and 2nd Battalions in the Peninsula. These battalions were removed to America, and returned in time to be present at Waterloo. The 3rd Battalion served throughout the Peninsular campaigns, unlike its senior battalions, but was disembodied before that fearful battle.

SECOND BATTALION.

This battalion was raised in Ireland in 1805, head-quarters at Glasgow, where it was assembled. In January, 1806, it embarked and landed, and marched into Hanover, where it was cantoned for six weeks. Shortly after it returned to England.

SICILY.

In May, 1806, it embarked and sailed for the Mediterranean, landed in Sicily, where the 1st Battalion of the 27th was completed from the 2nd; and that battalion ordered to Malta, there to receive 300 recruits, mostly lads from seventeen to nineteen years of age. Malta was deemed an excellent garrison for health and instruction. Major Reeves, of the 3rd Battalion, received an order from the Horse Guards directing him to repair to Malta and take the command of the 2nd Battalion 27th.

Major-General Oaks, in 1808, having made a favourable report of the state and discipline of the 2nd Battalion 27th, Sir John Stuart ordered it to Sicily. The regiment embarked with the army at Malazzo in June, 1809, for the Bay of Naples; disembarked on the islands of Procida and Ischia, which were easily taken. Shortly after returned to Sicily, and moved to Palermo.

On the 12th of December, 1812, the commander of the forces, Lord William Bentinck, inspected the battalion, and observing that a drummer in each company carried a bugle, inquired if the regiment was instructed in light movements —then not usual. Being answered in the affirmative, and those movements being executed to his satisfaction, he inquired if the regiment could be ready to embark next day for Spain. He was answered it was ready to embark at that moment, as it was in marching order. However, next day was fixed, the 13th of December. It embarked in men-of-war and sailed the same day for Spain, and landed at Alicante on the 3rd of January, and was marched off to the advance posts, where it joined the Light Division, consisting of the light and rifle companies of the king's German Legion, the Calabrese Free Corps, a brigade of mounted guns, carried on the backs of mules, and a squadron of hussars formed of deserters from the French, and consisting of Hungarians, Poles, &c.

ALCOY.

On the 6th of March the Light Division made a reconnaissance as far as the town of Alcoy, and drove the enemy out of it. In this skirmish Captain Parsons, of the light company, and sixteen rank and file were wounded, three of them mortally. Captain Parsons afterwards died of his wounds.

ALSAFARA.

On the 15th March the grenadier company of the 2nd Battalion had an opportunity of distinguishing themselves in a reconnoitring party under Major-General Donkin, by a spirited attack upon about 300 of the enemy, and driving them from the village of Alsafara, of which they took possession. On this occasion, Captain Waldron, Lieutenants Manly and Talbot, with the non-commissioned officers and soldiers, were thanked in general orders, and the gallant conduct of the company, who with Captain Waldron had broken through a battalion of the enemy, was mentioned in the public despatches of the general commanding.

BIAR.

The Light Division moved to the front, and occupied the village of Biar. On the 11th of April a confirmation was received of Marshal Suchet's advance with his whole force. Shortly after, their light troops took possession of Vilhena, and made prisoners a fine-looking Spanish battalion of Cascadores, absurdly left there by General Elio without provisions in the castle. Early in the morning of the 12th the enemy moved out of Vilhena into the plain—cavalry, artillery, and infantry; and, as no further doubt remained as to his intention, Colonel Adam, the commander of the Light Division, made the necessary disposition for his reception. The division now consisted of the 2nd 27th, the light and rifles of the German Legion, the Calabrese Free Corps, the 1st Italian Levy, one foreign troop of cavalry, and the mountain guns on the mules (neither useful nor ornamental: they were a positive nuisance).

The principal street in Biar, and a convent that commanded it, and a road at the back of the village were occupied by part of the Calabrese Free Corps, the eighth company of the 27th (picked shots), and a company of the German Rifles. This was thought sufficient, as it was truly conjectured, when the enemy found such resistance in the streets, they would burn the village. However, they suffered considerably in the streets. As soon as it was perceived they were gaining ground on the flanks of the village, the signal was sounded to those in the town to retire behind the line formed in rear of the place. This was smartly executed. After some arrangement and hesitation, the enemy debouched from the village, covered by a cloud of tirailleurs.

These were met and well punished by the German riflemen and the light company of the 27th, and driven back on their columns, while the Calabrese and the Italians kept up a heavy fire. Notwithstanding, they deployed and moved forward in line, supported by several heavy columns. Colonel Adam's orders were to fall back gradually, but to dispute the ground where practicable. They therefore retrograded to a new position. The 27th executed this movement by wings, the riflemen and light infantry thumping the enemy's tirailleurs handsomely. At this period of the action Colonel Adam was wounded. . . . . The action became rather warm, when a body of cuirassiers was observed gaining the road in front of the left of the 27th. Lieutenant-Colonel Reeve, recognizing their intentions, detached Captain Hare with three companies to take post behind some rocks where the road made an angle, and there wait until those cuirassiers had committed themselves, which they soon did, galloping down the road, and calling to the 27th—"*Bas vos armes!*" At this juncture they received a volley from the ambuscade that made them reel, and those that could galloped back, leaving the best part of them kicking in the road. It was remarked that the musket balls passed through cuirass and man. In fact the muskets almost touched some of them. This had a good effect, as the line formed in front of the 27th fell back, throwing out sharpshooters. Here was an opportunity for a charge. But the 27th had orders to fall back, and did so in ordinary time. The last check made them much less pressing, and they brought up several guns. Under this fire the columns of attack again moved forward, as usual covered by sharpshooters, and as usual were met by our light troops. Lieutenants Manly, Duhigg, and Jameson, of the 27th, were by this time wounded, all severely, and several of our men; but the enemy fell very thickly. Colonel Adam having had his wound dressed, came up, and gave orders to fall back gently, which was done. The enemy seemed glad to get rid of us, and after five hours' fighting we marched into our position on the left of the army at Castalla. Marshal Suchet took up his position. His army was said to be the first in Spain for their numbers, and their appearance did not contradict this. They were perfect in all arms, but particularly a magnificent cavalry, in which arm we were deficient. The baggage was ordered off, and all prepared for a general action, and the morrow was anxiously looked for.

CASTALLA.

On the 13th April the enemy appeared in order of battle, and in three divisions attacked the left of the combined line in position at Castalla, under Marshal Suchet in person. The 2nd Battalion 27th particularly distinguished itself in a brilliant charge with the bayonet upon a column of French grenadiers, who were selected by Marshal Suchet to attack that part of the position. The despatch of Sir John Murray stated:—"The most gallant charge of the 2nd Battalion, 27th Regiment, led by Colonel Adam and Lieutenant-Colonel Reeves, decided the fate of the day."

Sir William Napier, in speaking of this battle, says, "This exploit was erroneously attributed to Colonel Adam, but it was ordered and conducted by Colonel Reeves alone."* Sir William Napier details the account of this combat. It is almost word for word the same as the description given of it in "Wars of the Nineteenth Century," book iv., p. 39:—

"But against Adam the French grenadiers mounted the hill resolutely, when they were encountered by the 27th Regiment in a terrible crash of battle. Here a singular and unprecedented event happened. As the French were deploying their columns, a French grenadier officer, advancing to the front alone, challenged any English officer to single combat: for from the nature of the ground the troops stood close to one another. Captain Waldron, of the 27th, an agile Irishman of boiling courage, instantly leapt forward to accept the duel, and the hostile lines looked on without firing a shot, while the swords of the two champions clashed and glittered in the sun; but Waldron cleft his adversary's head in twain in the very first encounter, and

* See Appendix.

the 27th Regiment, springing up from the ground with a deafening shout, led by their colonel, Reeves, now charged and sent the enemy, *malgré* their numbers and courage, down the mountain side in great confusion. For this gallant conduct Captain Waldron obtained the brevet rank of major."

The Frenchman's sword was a sabre of honour given him by Buonaparte, which was seized by the victor and sent by the quarter-master-general to his Royal Highness the Duke of York, who, ever alive to deeds of heroism, had Captain Waldron gazetted a brevet-major. Captain Hare, of the 27th, for his conduct on the 12th, when the cuirassiers got a dressing (luckily the occurrence was observed by the commander of the forces), got the brevet rank of major; and Brevet-Lieutenant-Colonel Reeves got the effective lieutenant-colonelcy of the regiment, over the senior major.

A ludicrous circumstance occurred while the French were advancing, just before the charge. From the scarcity of money in the military chest the regiment was literally without a sixpence. A dead silence prevailed, the men lying down, when a fellow put up his head, and with characteristic drollery exclaimed, "By J—s, it's our money they want, boys!" This produced a roar of laughter, and was soon followed by "Cap," "Fire," and "Charge."

The French retired during the night, leaving their watch-fires burning, and were gone some time before it was observed. Their loss in the two days was estimated as at least 4000 men. We lost several of the men belonging to the light and rifle companies of the German Legion, and three of their officers. Better or braver soldiers never drew trigger, and the men of the 27th regretted them much, on taking up their old quarters at Biar on the 15th. In the two days' fighting the 27th had three subalterns severely wounded, two subalterns slightly, and fifty-three rank and file killed, wounded, and missing.

The following is an extract from a private letter from Lieutenant-General Sir John Murray to Colonel Torrens, dated Alicante, June 30th, 1813:—

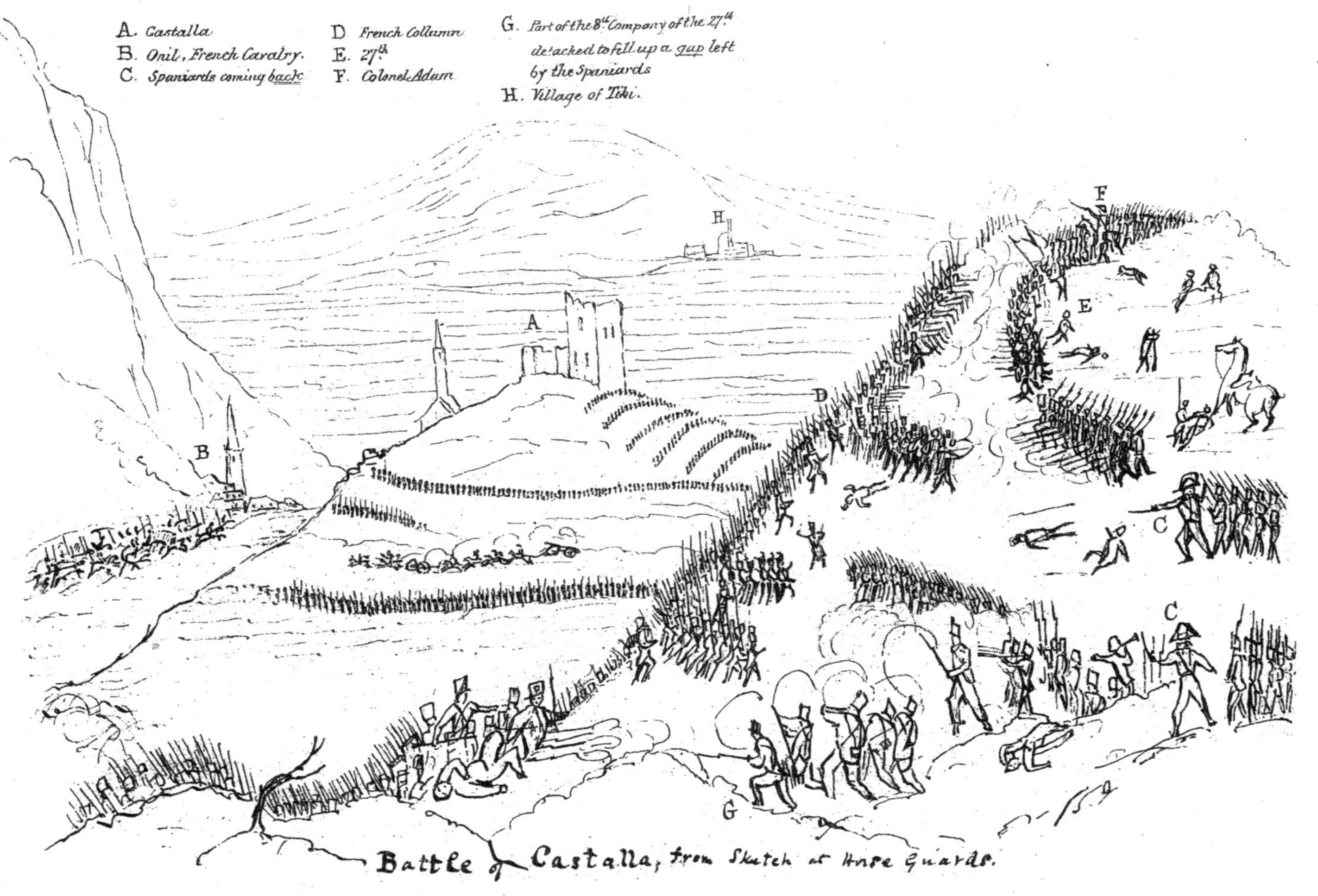

Battle of Castalla, from Sketch at Horse Guards.

"I hear Colonel Burke and Major Carey are to be lieutenant-colonels. This rank is certainly well bestowed, but there is an officer who, on the 12th and 13th April, distinguished himself pre-eminently, who is omitted. I mean Lieutenant-Colonel Reeves, of the 2nd-27th. He is a most gallant and excellent officer, and if anything could possibly be done for him, I believe the whole army here would be gratified, and nobody more than myself. That regiment had the brunt of the business on both days, and Lieutenant-Colonel Reeves conducted it with the greatest gallantry and skill."

TARRAGONA.

In May, the army under Sir John Murray embarked at Alicante for Tarragona on the 31st. Having landed near Tarragona on the 2nd June, the siege of that fortress commenced immediately, and continued till the evening of the 12th when it was broken up in consequence of Marshal Suchet moving with a considerable force for its relief. During the siege, the advance under Colonel Adam, and the forming part of the covering army, the loss of the 27th was one officer (Captain Nethery) killed, and a few privates. After the unsuccessful siege of Tarragona, the army returned to Alicante, but was soon again in motion under Lord William Bentinck, who had assumed the command, following Marshal Suchet in his retreat from Valencia to Barcelona.

ORDELL.

The advance under Colonel Adam was posted in the pass of Ordell, about fifteen miles in front of the main body of the army, which was at Villa Franca. The French army attacked and stormed that part at midnight of the 12th September. The troops fought desperately, but their position was at last carried. The enemy being repulsed in every attack which they attempted in front, at last succeeded in turning their flanks. Colonel Adam received two wounds, and at about one o'clock in the morning was obliged to leave the field. Lieutenant-Colonel Reeves succeeded in command. He was soon shot through the body, and also obliged to leave the field. Captain Mills then took the command, and was also desperately wounded through the body. The loss chiefly fell on the 27th Regiment, being the only British regiment engaged in this affair, having had eight officers and three hundred non-commissioned officers and men killed and wounded. Adjutant Taylor was killed. The following officers were wounded:—Lieutenant-Colonel Reeves, Captain Mills, Captain Winser, Lieutenant Talbot, Lieutenant M'Pherson, Lieutenant Manly, and Lieutenant Drewe.

On September 14th Marshal Suchet attacked the allied army in position at Villa Franca, which retired upon Arbos. The 1st Battalion 27th lost a few men on this day. The army returned into cantonment in Tarragona, Vandrills, and Arbos.

BLOCKADE OF BARCELONA.

1814.—On 14th February the blockade of Barcelona commenced, in which the 1st and 2nd Battalions were employed under Sir William Clinton, and continued until April. On April 14th broke up from Barcelona, a large proportion of this army being ordered to join the Duke of Wellington on the Garonne.

THE THREE BATTALIONS—1ST AND 3RD TO AMERICA.

1814.—May 24th arrived at Bordeaux, where the three battalions of the 27th met *for the first time.* The men of the 3rd Battalion were handed over to the 1st Battalion. The 1st and 3rd were ordered to America, and the 2nd to Cork. In August the 1st and 3rd Battalions arrived in Canada; the 1st was placed in brigade under Sir Manly

Power. The 3rd Battalion and 76th Regiment formed a demi-brigade under Major-General Sir Frederick Robinson, and which formed the advance of the army at the advance to Platsburgh. After the retreat from Platsburgh the 1st Battalion returned to Montreal, the 3rd to Chambly.

FROM AMERICA TO ENGLAND.

November.—Major-General Sir Manly Power having received orders to proceed with one battalion to reinforce the army at New Orleans, the major-general *selected* the 1st Battalion 27th from his own brigade to accompany him, and that battalion embarked on the 5th November at Montreal; 10th December, arrived at Halifax; 18th, sailed; January, Bermudas; 18th February, sailed; March, Port Royal, Jamaica; on 5th sailed; on 14th anchored at the entrance of the Mississippi; on 18th sailed; April 3rd, Jamaica; on 8th sailed for England. On May 10th, 1815, seven companies of the 1st Battalion arrived at Portsmouth, the remaining three companies (the head-quarters of the regiment) having separated in a gale of wind; the seven companies were disembarked, and joined the 2nd Battalion at Gosport barracks. Those companies being completed to 100 rank and file each, by drafts from the 2nd Battalion, they again embarked on the 17th for Ostende.

After the capitulation of Paris to the allied forces, Napoleon retired to the Island of Elba, and the Bourbons were restored to the monarchy. He instituted intrigues in his favour, and a constant correspondence was maintained with his relations and adherents. A conspiracy was formed, and eventually, on the 1st of March, 1815, he landed in the Gulf of San Juan. He was greeted with acclamations on all sides, soldiers flocked to his banner, and on the 9th he entered the Tuileries amidst a splendid array. Europe was again alarmed. Wellington was despatched to the continent, and Prussia sent forward her forces.

On May 24th the battalion disembarked at Ostende, conveyed to Ghent in boats, and there cantoned; on June

16th marched from Ghent; on 17th passed through Brussels without halting, and arrived on the morning of the 18th on the field of Waterloo.

WATERLOO.

On the memorable day of the 18th June the 6th Brigade, under Major-General Sir John Lambert, was left at the commencement of the action for some time in reserve in front of the village of Waterloo. The furious and repeated attacks of the enemy on both flanks having been most gallantly repulsed by the Guards and General Picton's division, their most desperate efforts were directed to break through the centre of the British line. In one of these they carried the farmhouse of La Haye Sainte. About eleven o'clock the 6th Brigade was brought up, and took its post in the centre of the first line, within one hundred yards or less of the farmhouse of La Haye Sainte, which the enemy then maintained possession of during the day, being supported by a powerful body of cavalry, who took advantage of the cover afforded them by the farmhouse and high walls that surrounded it, and there remained during the battle, watching every favourable opportunity to charge. The 27th was placed on the right of the brigade, having its right resting close to the Charleroi road. In this position the brigade continued during the action in squares of battalions, receiving and gallantly repulsing repeated and furious charges from the enemy's cuirassiers. On this trying occasion the intrepidity and discipline of the 27th remained unshaken, eager to advance, but restrained till it could be done with success, the difficulties of attack upon the farmhouse appearing insurmountable in the presence of a numerous body of cavalry.

In this manner the regiment stood with perfect firmness for many hours, exposed to a most galling and destructive fire, without having it in their power to make the smallest resistance or to return a shot, while the French had leisure to take cool aim. This trial of courage and of discipline was great, and might have shaken the firmness of the best troops; but the Enniskilleners supported it with the most perfect steadiness, and not a man moved from his post, except to fill up the vacancies of the dead and wounded. Two staff officers during the battle rode into the square of the 27th Regiment. Those officers were sent round by the Duke of Wellington to acquaint the troops that, sooner than yield a foot of ground to the enemy, his grace expected that every British soldier would prefer dying on the spot on which he then stood. Each of those officers was received by the soldiers of the 27th with three cheers, and the 27th well obeyed their illustrious chief's orders—as will be seen by the following return of killed and wounded in the square:—

	Killed.	Wounded.	Total.
Officers	2	13	478
Rank and file ...	103	360	

There are few, if any, instances in the military history of any nation, of so many men having fallen on so limited an extent of ground. The stubbornness with which the 27th maintained its ground against repeated charges of cavalry, and exposed for hours to most incessant and destructive fire, was a trial of character of the British soldier, to which the foreign officers who witnessed it, and many of the enemy's officers also, were heard to have done ample justice.

"The Irish howl set up by the Inniskilling and other Irish regiments is reported to have carried almost as much dismay into the ranks of their enemy as their swords. The stubborn bravery and conduct of these regiments contributed much to the success of the day; it having been their lot to find themselves in the hottest part of the action, innumerable opportunities were afforded them of showing their devotion to their country's honour and exalted sense of gallantry and duty."

During this battle, when the firing had continued for a long time, and numbers of the men had fallen, the ammunition of the survivors was expended, and Sergeant Scott was ordered to take a few men and fetch, "if God spared them," a supply of ammunition from the waggons. Scott, with great presence of mind and coolness, suggested, "Why not the dead supply the living?" The hint was taken and acted on, and the cartouches of the numerous dead, dying, and wounded afforded a more than sufficient supply for the remainder of the day. The 27th having had to make a forced march of some thirty miles the day before the battle, the men carried their knapsack, blanket, &c., and Scott had slung round his neck a tin kettle, which was tilted up by the knapsack, so as to expose it to the enemy's bullets, five of which struck it in rapid succession, thereby making a considerable jingling, reminding the sergeant so forcibly of a tinker at work, that he cried out, "D— y—." An officer near him, surprised at this strange soliloquy at such a time, asked whom on earth was he cursing. "The damned tinker," replied Scott, "that's playing away at my tin kettle."

The following is a list of officers of the regiment present at Waterloo:—

Captain J. Hare, major commanding. Wounded.
„ John Tucker. Wounded.
„ G. Holmes. Killed.
Lieut. G. Macdonald. Wounded.
„ W. Henderson. Wounded.
„ Richard Handcock. Wounded.
„ E. W. Drewe. Wounded.
„ John Betty.*
„ W. F. Fortescue. Wounded.
„ W. Talbot.*
Lieut. John Millar. Wounded.
„ Charles Manley. Wounded.
„ Thos. Craddock. Wounded.
Ensign T. Handcock. Wounded.
„ R. Ireland. Killed.
„ Thos. Smith. Wounded.
„ John Ditmar. Wounded.
Quarter-Master T. Taylor.
Assistant-Surgeon Thos. Mostyn.
„ G. Fitz Gerald.

The Emperor of Russia conferred the order of St. Wladimir, upon Major Hare, in testimony of his services and bravery, particularly at the battle of Waterloo.

John Mills, in his "Corporal Crump's Narrative," says:— "It was sorely trying to our men to await the charges of the enemy, hour after hour, each square standing on its appointed ground, and as the gaps were made in the front ranks, for others to step forward to supply their places, only to meet with the same fate. It was almost more than could be expected from any mortal troops; for it is one

* Either Lieutenant Betty or Lieutenant Talbot was not present, but according to the returns I cannot ascertain which.

thing to rush forward cheering and shouting in a charge, and quite another to receive it in cold blood, with your comrades dropping around you like hail. As an instance of what British soldiers can take, as well as give, I may mention that the 27th Regiment had 400 men and every officer, except one subaltern, knocked over in square, neither moving an inch nor pulling a trigger. Many a chicken heart can be roused to do a bold act; but the true courage of a soldier shows itself in deeds such as these."

The heroism of the 27th at Waterloo has been wedded to romance. Our gifted countryman, Charles Lever, has added to the charms of fiction the beauties of history. In his famous and deservedly popular novel, "Charles O'Malley," after relating the story of the Peninsular War with all the grace and eloquence of his language, he comes to Waterloo. He speaks with pride of the Enniskilleners (dragoons) and "the flashing features" of "my own island countrymen." Further on he comes to speak of the 27th Inniskilling Foot. I will quote his words:—

"By a slight circuitous route we reached the road, upon which a mass of dismounted artillery, carts, baggage waggons, and tumbrils were heaped together as a barricade against the attack of the French dragoons, who more than once had penetrated to the very crest of our position. Close to this and on a little rising ground, from which a view of the entire field extended, from Hougomont to the far left, the Duke of Wellington stood, surrounded by his staff. His eye was bent upon the valley before him, where the advancing columns of Ney's attack still pressed onwards; while the fire of sixty great guns poured death and carnage into his lines. The second Belgian division, routed and broken, had fallen back upon the 27th Regiment, who had merely time to throw themselves into square, when Milhaud's cuirassiers, armed with their terrible, long, straight swords, came sweeping upon them. A LINE OF IMPASSABLE BAYONETS, A LIVING *chevaux-de-frise* OF THE BEST BLOOD OF BRITAIN, STOOD FIRM AND MOTIONLESS BEFORE THE SHOCK. The French *mitraille* played mercilessly on the ranks, but the chasms were filled up like magic, and in vain the bold horsemen of Gaul galloped round the bristling files. At length the word 'Fire!' was heard within the square, and, as the bullets at pistol-range rattled upon them, the cuirass afforded them no defence against the deadly volley. Men and horses rolled indiscriminately upon the earth."

A British officer, who was an eye-witness of the gallant conduct of the 27th, says, "If ever the sovereign give them another motto, it should be *Muzzle to Muzzle,* for so they fought at Waterloo." *

Lieutenant-Colonel (now Major-General) Charles Beckworth, one of the oldest and most distinguished of the staff officers in the Peninsula, was at Waterloo, and thus described the conduct of the 1st Battalion of the 27th, only six hundred strong, and all young soldiers, who had never seen a shot fired before:—

"I have seen many proofs of the intrepidity of British soldiers, but the conduct of the 1st Battalion of the 27th at Waterloo *was the most extraordinary* I ever beheld. They were placed in column, under the fire of several battalions, and kept there without forming a line, or firing a shot in return. The bullets ploughed through their mass until nearly two-thirds went down, and the survivors, as if fearing their own nerves would give way, but resolved not to go back, pressed their heads and shoulders inwards, forming a solid ring; and thus leaning together and striving as it were to push all to the centre, moved round and round, stamping like men in a mill; apparently frantic with horror and excitement, yet firm in their resolution to overcome nature and fall to the last man, continued their tramp, constantly closing in, etc., lessening their glorious ring as the crashing flights of metal continued to tear through the living mass."

"A private of the 27th, wounded very severely, was carried off the field of battle by his wife, then far advanced in pregnancy. She, too, was severely wounded by a shell, and both of them lay a long while in one of the hospitals in Antwerp in a hopeless state. The poor man has lost both his arms,

* "History of Waterloo," by Sergeant-Major Cotton.

the woman extremely lame, and given birth to a daughter, to which the Duke of York, it is said, has stood godfather, by the name of Frederica McMullen Waterloo." *

PARIS.

On the day following the battle of Waterloo the regiment marched in pursuit of the enemy. On the 6th of July they entered Paris. Sir Lowry Cole's division encamped at Neuilly. On October 28th the 1st Battalion broke from cantonment at Maison, thence to St. Germains, where the 1st Battalion was completed to 1000 strong by drafts from the 3rd Battalion. Afterwards marched from Paris to Cambray, where the regiment was cantoned with the army of occupation (in 1816) in France.

Before the regiment proceeded to Calais for embarkation, Brigadier-General Sir John Lambert said he wished it to be noted in the Regimental Records the high opinion he entertained of the gallantry of the 27th, and their good conduct in the field, which was especially marked in the memorable battle of Waterloo.

HOME.

On 6th March, 1817, the regiment embarked at Calais, after an absence of twenty-four years of unremitting and active foreign service in every quarter where a British army has been employed, and in the course of the numerous trials of courage and of discipline to which three battalions of the 27th were exposed in many a hard-fought battle and desperate conflict with the enemy. Yet on no occasion has any of them failed to evince the same gallant spirit for which the Enniskillen regiment has ever been conspicuous. No battalion was ever more actively employed than the 3rd Battalion 27th from 1808 to the termination of the Peninsular War, having during that period formed a battalion of the brave 4th Division, commanded by their distinguished colonel and Enniskillener, Sir Lowry Cole.

* Anecdotes of Waterloo.

PARLIAMENT'S THANKS TO SIR LOWRY COLE.

At this time Sir Lowry was member of Parliament for Enniskillen. On the 20th of July, 1816, Lieutenant-General Sir Lowry Cole was again addressed in his place by the Speaker of the House of Commons, as follows:—"In that victorious army which re-established the thrones of our allies, though all were brave, yet by the fortune of war the skill and valour of some were rendered conspicuous above the rest, and the gallant 4th Division was distinguished throughout by the highest praise for its enthusiastic courage and heroic devotion. Of that division, to whom all eyes were turned, in every battle you, sir, had the chief command; and your growing renown was well supported by many brave companions in arms, whose names will ever live in our annals."

THIRD BATTALION.

The 27th Regiment was only composed of one battalion till 1804, when the unbounded ambition of Buonaparte, and his demonstration of invading England, rendered it necessary to increase the army considerably, by raising a second battalion to each regiment of the line. The 27th was at this time in Sicily; yet the corps was so popular in Ireland, that its second battalion was completed in less than twelve months from its formation, from men raised in Ireland under the Defence Act; and had so many supernumeraries, that in less than twelve months more there were sufficient to form a third battalion, which was embodied in Edinburgh on the 25th of September, 1805; and being composed chiefly of young men and lads, it was kept at home as a depôt for recruiting, and for supplying the 1st and 2nd Battalions with men. The 2nd Battalion, which was embodied at Glasgow the preceding year, was also composed of men raised in a similar manner. Both battalions had permission to extend their services and receive a bounty. The establishment was first fixed at 400 rank and file, afterwards increased to 600, 800, and 1000; and ultimately each of the three battalions 27th Regiment was increased to the establishment of 1200 rank and file, with a recruiting company to each.

In October, 1805, the 3rd Battalion was removed from Edinburgh to Dunbar. On January 7th, 1806, the battalion returned to Edinburgh Castle; on June 15th removed to Stirling Castle; August 6th, to Glasgow barracks; October, to Ayr barracks, where the limited service men who declined to extend their services, about 300 in number, were transferred to the 14th Garrison Battalion, and the unlimited men, upwards of 400, were sent to the 1st and 2nd Battalions of the regiment. The officers and non-commissioned officers of the 3rd Battalion being ordered to Ireland to recruit, they were stationed in Belfast in December. In February, 1807, they were removed to Omagh, the establishment being here soon completed by the recruiting parties, and volunteers from the Irish militia.

ENNISKILLEN.

In July Enniskillen had this battalion of its own regiment, where it remained till the May following. In 1808 marched from Enniskillen to the camp which was forming at the Curragh, Kildare, under Lieutenant-General Sir David Baird. On July 18th broke up from the Curragh and proceeded to Middletown barracks, Co. Cork, to be ready for embarkation, where it left 300 boys as a depôt.

PORTUGAL.

On September 9th the battalion embarked at Cove, 800 strong, under the command of Lieutenant-Colonel Maclean, with a division of troops under Sir David Baird, consisting of about 3000 infantry, for Falmouth, where being reinforced by 5000 more troops, the fleet sailed from Falmouth on the 9th October, and arrived at Corunna on the 14th of the same month. On the 25th the 3rd Battalion sailed for Lisbon, where it disembarked on the 2nd November. On January 14th embarked at Lisbon with some other troops, for Cadiz, under the command of Major-General R. Mackenzie. On February 11th the fleet with the troops from Cadiz returned to Lisbon, accounts having been received of the battle of Corunna and of the embarkation of the British troops, the grenadier company of the 27th being the only troops permitted to land at Cadiz to form a funeral procession in removing the body of Lieutenant-Colonel Sir George Smith, to be embarked for interment at Gibraltar. On March 14th disembarked at Beleni, and occupied barracks near Lisbon.

Lieutenant-General Sir John Craddock, who had succeeded to the command of the troops in Portugal after the departure of Sir John Moore, on receiving reinforcements from England, made preparations for advancing on Oporto, in order to dislodge the French under Marshal Soult. On April 8th the army broke up from their defensive position at Lisbon, amounting to about 18,000 men, and on the 22nd arrived at the city of Leyria. On the 2nd of May Sir Arthur Wellesley arrived at Coimbra. The whole of the army was now assembled. Major-General Mackenzie's brigade and the Lusitanian legion had advanced to Abrantes, on the Tagus. This division was advanced with a view to check the advance of Marshal Victor into Portugal, who was then in the neighbourhood of Badajoz. The men of the 3rd Battalion 27th having contracted a severe fever from their long stay on board ship, returned to Lisbon in June, and when recovered returned to Badajoz on the 8th October, 1809. On December 29th commenced the march with the army to the north of Portugal, and arrived at Celorico on 17th January, 1810.

SPAIN.

The battalion marched from Celorico to Guarda, where Sir Lowry Cole's head-quarters of the 4th Division were stationed, which then became composed of the following regiments:—7th, 11th, 27th (under Colonel Maclean), 40th, 48th, 97th; and 11th and 23rd Portuguese. In this position the army remained undisturbed until September, when, in consequence of the surrender of Ciudad Rodrigo by the Spaniards and of Almeida by the Portuguese to the enemy, Marshal Massena crossed the Mondego with his whole army on the 18th, 19th, and 20th September at the bridge of Torons, below Celorico. This movement of the enemy was met by the Duke of Wellington, who placed his army on the right bank of the Mondego, and retired upon the position at Busaco, Sir Lowry Cole's division by Panhouses and St. Maria.

BUSACO.

September 26th.—Before daybreak the several divisions of British and Portuguese troops began to ascend the heights of Busaco, from whence the whole of the enemy's force, amounting to not less than 60,000 infantry and a very heavy body of cavalry, was distinctly discerned. That evening an affair of picquets took place. Next morning, the 27th September, the enemy made two desperate attacks upon the right and centre of the allied army, and were foiled in both. On the 29th the army recrossed the Mondego, and continued to retreat upon the position in front of Lisburn, with the right at Alhondra on the Tagus, passing Torres Vedras, and the left on the Lee, the 4th Division at Patameira. Both armies remained inactive in their positions until the 14th November. The enemy having that day returned to Santarem, Lord Wellington followed him on the 15th, and fixed his quarters at Cartaxe, the 4th Division at Azambija and Virtudas. Here both armies remained again quiet for some months.

REDINHA.

At the affair at Redinha on the 12th March, 1811, the 4th Division was led to attack the French by General

Spencer, who took post in front of the 27th; and he, whilst the troops were advancing in line under a heavy fire of artillery, took occasion to declare that he had never seen troops march to the enemy in better style than this young battalion. This was only the second occasion on which the battalion was engaged.

OLIVENCA.

The 27th was also present at Olivença. The 4th Division then invested Fort St. Christoval. During a sortie from Badajoz the following officers were killed:—Major Berningham, Captain Smith, and Ensign M'Cord. Officers wounded: Lieutenant-Colonel Maclean, Captain Pring, and Lieutenants Gordon, Dobbin, and Levinge. On Marshal Soult advancing from Seville, the 4th Division recrossed the Guadiana, and pursued the enemy after the battle of Albuera. Meanwhile, the French Marshal Soult held Badajoz.

ALDEA DE PONTE.

Marshal Marmont having crossed the Tagus and established himself at Placentia in the beginning of August, the main body of the British army made a parallel movement by crossing the river at Villa Velha, the 5th Division proceeding by Albuquerque and Fuente de Geralde. On the 26th September the enemy advanced from Ciudad Rodrigo, and the 4th Division retreated to Aldea de Ponte, where it had an affair with the enemy on the 27th September.

CIUDAD RODRIGO.

1812.—The enemy having retreated, pursued by the 4th Division, to Ciudad Rodrigo, the division took up winter quarters, the 27th at Almeida near Ciudad Rodrigo. The 27th having marched to Coimbra for clothing and necessaries, they returned to Almeida in time to take part in the siege of Ciudad Rodrigo from the 9th to the 19th January, 1812, when it surrendered. After the capture of Ciudad Rodrigo the 27th occupied Villa de Corvo.

BADAJOZ.

In March the 27th proceeded to Badajoz. On the 16th, Badajoz was invested by the 3rd, 4th, and Light Divisions of infantry on the left bank of the Guadiana. On the following day the troops broke ground and established a parallel within 300 yards of an outwork called the Picorona. The operations of the siege were continued, notwithstanding most severe weather, from the 20th to the 28th of March. On the latter day 28 pieces of ordnance were opened from two batteries; and La Picorona was carried in the evening by Sir James Kempt's division. Practicable breaches being effected in the evening of the 6th, Lord Wellington determined on an immediate assault of the fortress.

The attack was made at ten o'clock at night, the 4th and Light Divisions moving on to the attack along the left of the river Rivillas and of the inundation. They were not perceived by the enemy until they reached the covered way, and the advanced guards of the two divisions descended into the ditch protected by the fire of parties stationed on the glacis for that purpose, and they advanced to the assault of the breaches with the utmost intrepidity; but such was the nature of the obstacles prepared on the top and behind the breaches, and so determined the enemy's resistance, that the British troops could not establish themselves within the place. These attempts were repeated till after midnight, with the loss of many brave officers and soldiers, when Lord Wellington, finding that success was not to be attained, ordered the 4th and Light Divisions to retreat to the ground on which they had assembled for the attack.

In the mean time General Walker's brigade, which was intended to have made a false attack on the left, finding but little resistance (the enemy having brought all their forces to the breaches), escaladed the bastion of St. Vincente. In consequence of this success all resistance ceased, and at daylight General Phillipon surrendered the fortress. The 27th had one captain, four subalterns killed; two field officers, nine subalterns wounded: viz. Captain Jones, and Lieutenants Levinge, Simcoe, White, and Ensign Warrington, killed; and wounded—Major Knight, lieutenant-colonel; Brevet-Lieutenant-Colonel Ward; Lieutenants Gordon, Pollock, Thornton, Wier, Moore, Radcliffe, Hanly, and Phillips; and Adjutant Davidson. There were three sergeants, 45 rank and file killed; nine sergeants, 225 rank and file wounded. Although every man performed well his duty on this occasion, yet it would be an injustice to the memory of two brave non-commissioned officers, in recording the gallant deeds of their corps, to pass unnoticed the very gallant conduct of Sergeant-Major Groove and Sergeant Carlisle, who (their officers being all either killed and wounded) were both killed in the breach, bravely encouraging their men.

April 27th.—Broke up from Badajoz for the north of Portugal, crossed the Tagus at Villa Velha, and arrived at St. Joan de Pesquire early in May.

June 5th.—Advanced into Spain by Penhel, Almeida, Ciudad Rodrigo, and the banks of the Tormes near Salamanca.

On the 17th Lord Wellington, with the main body of the combined army, crossed the Tormes at Salamanca, which the enemy abandoned, leaving a garrison of 1000 men in a fortified convent on the brink of the river, and commanding the bridge. This was taken by the 6th Division on the 27th. After several ineffectual attempts made to relieve the garrison, Marmont, being thus defeated in his object of relieving the garrison of Salamanca, withdrew his army behind the Douro, destroying the bridge, and followed by the combined army.

On the 16th of July a considerable body of the enemy crossed the Douro at Toro. These troops recrossed the river on the same night at the same place; moved to Fordesillas, and again crossed the Douro on the morning of the 17th, on which day Marmont assembled the whole of his force at Novo del Rey.

CARNIZAL.

On the morning of the 18th the enemy attacked the 4th Division at Carnizal. The 27th and 40th Regiments made a gallant charge upon a strong column of the enemy who tried to turn the left of the British army. Although the enemy were double the number of these two corps, yet they were put to flight in this affair. The 27th had two lieutenants (Lieutenants Radcliff and Davidson) and eleven rank and file killed; one sergeant and fifty-seven rank and file wounded.

SALAMANCA.

On the 21st crossed the Tormes near Salamanca; and on the 22nd the 27th was ordered to take position on the top on the right of the British line, the post of the Anapiles, and the key of the position, and of so much importance that Lord Wellington in person addressed the commanding officer, Colonel Maclean, in the following words:—"You must defend this position as long as you have a man." The regiment kept possession of the position during the action, and until the British army obtained so decisive a victory that the action at last became more a pursuit than a fight. The 27th had one officer (Lieutenant Gordon) and seventeen rank and file wounded.

MADRID.

The victorious army continued its march after the enemy to Alba de Tormes, Segovia, and entered Madrid on the 12th of August. The troops were received in the Spanish capital with the most enthusiastic joy. The inhabitants met them several miles from the city. The houses were decorated, and illuminated by night.

After remaining a few days at Madrid, the 27th marched with the 4th Division to Escurial, five leagues from the city, and were cantoned in the villages adjacent. October 5th, broke up from cantonments near Escurial and marched to Valledemona, and joined Lieutenant-General Lord Hill's corps near Aranjuez. The French army under Soult, having raised the siege of Cadiz, advanced on Madrid and crossed the Tagus near Aranjuez. Lieutenant-General Hill therefore quitted his position on the Tacoma and marched northwards, and in the beginning of November this division of the army arrived unmolested on the Adaja, and continued their march to Alba de Tormes, where they arrived on the 12th November, and there met the troops from Burgos, with Lord Wellington, who had given up the siege of the place.

The combined army continued its retreat on the 15th to the frontiers of Portugal. The roads were bad, and owing to the inclemency of the weather at that late season of the year, together with the troops not having slept under cover for near a month, they suffered severely. Several men having died from fatigue and exposure to almost incessant wet, many were taken prisoners from falling behind, being unable to keep up with the columns. The 27th lost only four men during this severe and fatiguing march, and it was generally remarked that the young Enniskilleners (as this battalion was always called in the army, being for the greater part young, growing lads) marched better than any corps on that service.

On the 24th of November the head-quarters of the allied army were again at Frezuada, on the Portuguese frontier. The greater part of the enemy's forces had recrossed the Tormes, and were marching towards the Douro. The army went into winter quarters, the 27th to Villa de Corvo near Alameida. Thus ended the campaign of 1812.

VITTORIA.

After various meetings, on the 14th and 15th June, 1813, the Marquis of Wellington crossed the Ebro with his army, and continued his march to Vittoria.

On the night of the 19th June the French army, commanded by Joseph Buonaparte, having Marshal Jourdan as his major-general, took up a position in front of Vittoria. Lord Wellington halted the allied army on the 20th, in order to close up its columns and to reconnoitre the enemy's position.

On the morning of the 21st the battle commenced by Sir Rowland Hill driving the enemy from the left of his position on the heights of La Puebla. The battle soon became general, and the enemy was driven at all points. Their retreat was so rapid that they were unable to draw off their baggage and artillery. 151 pieces of cannon, 415 waggons of ammunition, and all their baggage, provisions, cattle, and treasure, and a considerable number of prisoners, fell into the hands of the British. The battalion in this action had one officer (Lieutenant Gordon) killed, two officers (Lieutenants Wier and Hiler) and one volunteer (Murray) wounded; seven rank and file, three sergeants, and thirty rank and file wounded. The army pursued the enemy to Pamplona.

On the 25th the French army commenced to retreat from the neighbourhood of Pamplona by the road of Roncesvalles into France, followed by the light troops of the allies, and on the next day Pamplona was invested. The 27th remained at the blockade of that fortress until the middle of July, when it pursued the enemy to Roncesvalles.

RONCESVALLES.

On the 24th July Marshal Soult collected at St. Jean de Piedde Port the right and left wings of his army, and a division of his centre, with some cavalry, amounting in all to between 30,000 and 40,000; and on the 25th attacked Sir Lowry Cole's and Major-General Byng's divisions at Roncesvalles. The position was maintained during the day, but being turned in the evening it was necessary to abandon it in the night, and the division drew back to recover the blockade of Pamplona, the 27th covering the retreat of the 4th Division.

PYRENEES.

From the 27th to the 28th of July the enemy made several attacks on the position occupied by the 4th Division, the importance of which post rendered it an object of vigorous assault and defence, in all which they were gallantly repulsed. On the 28th a last and desperate effort was made to recover the division, every regiment of which charged with the bayonet several times, and it happened to the 27th to charge four different times. The result of this day's action was a repulse of the enemy with great slaughter. Sir Lowry Cole immediately pursued them, and engaged them again at Estevan. In those different actions in the Pyrenees and near Pamplona, from 25th July to 1st of August, 1813, the loss of the 27th was three officers (Captain White, Adjutant Burns, Lieutenant Crawford) killed, twelve officers (Lieutenant-Colonel Maclean; Captains Hamilton and Butler; Lieutenants Pratt, Pollock, and Hanley; Ensigns Drewe, Byrne, Radcliff, Owens, and Clowes; and Surgeon Wray) wounded. Fifty-seven rank and file were killed; 200 rank and file wounded.

The conduct of the battalion at the battle of the Pyrenees is thus mentioned in the Marquis of Wellington's despatches:—"The battle now became general along the whole front of the heights occupied by the 4th Division, and in every part in our favour, excepting where one battalion of the 10th Portuguese were posted. This battalion having been overpowered, and having been obliged to give way immediately on the right of Major-General Ross's brigade, the enemy established themselves on our line, and Major-General Ross was obliged to withdraw from his post. I, however, ordered the 27th and 48th Regiments to charge, first that body of the enemy which had first established themselves on the heights, and next those on the left. Both attacks succeeded, and the enemy were driven down with immense loss.'

ST. SEBASTIAN.

At the storming of St. Sebastian on 31st August, a detachment of the regiment, who, with Lieutenant Harding and volunteers Kenyon and Cranston, volunteered the forlorn hope, behaved very gallantly in the assault on the breach. Those three gentlemen were killed, with six rank and file, and two rank and file were wounded.

NIVELLE.

On the 10th November the regiment was engaged in the attack of the enemy in their fortified position at Nivelle. The operations began at daylight, and from the resistance experienced it was night when the army had effected their purpose of gaining the rear of the enemy's right. The French having quitted during the night all their works and posts in front of St. Jean de Luz, were pursued by the British troops across the Nivelle, who took fifty-one pieces of cannon, six tumbrils of ammunition, and 1400 prisoners. The 27th in this battle lost one field officer (Major Johnstone) killed; three officers wounded; one sergeant, eight rank and file killed; one sergeant and fifty rank and file wounded.

NIVE.

After the retreat from Nivelle the enemy occupied a very strong position in front of Bayonne, under the fire of that fortress, with posts on the rivers Adour and Nive.

On the 9th December the British army crossed the Nive at two points. A series of movements ensued on both sides during the 9th, 10th, 11th, 12th, and 13th December, connected with the passage of the Nive, in which the enemy made several desperate attacks upon part of the allied army, and which were all repelled with great gallantry, and the enemy withdrew to their entrenchments. The eventful campaign of 1813 being here terminated, Lord Wellington with his victorious army took up winter quarters upon French territory.

ORTHES.

1814.—The allied army crossed the Gave de Pau on the morning of the 25th February, and early on the 27th found the enemy in a strong position near Orthes. Marshal Beresford was ordered to attack the enemy's right with the 4th Division under Lieutenant-General Sir Lowry Cole, while Sir Rowland Hill was to attack and turn the left. After an obstinate resistance the 4th Division carried the village of St. Boes. A general attack being then ordered on the whole of the enemy's line, they were dislodged at all points, and were so closely pursued that their retreat became a flight, and they were thrown into the utmost confusion. The pursuit was continued till dark, and was resumed the next day to St. Sever, and on the 1st of March the army crossed the Adour. Loss of 27th—one officer (Lieutenant Nixon) wounded; one sergeant killed; 14 rank and file wounded. The battalion having been ordered to Bayonne for clothing, it arrived just in time to join the 1st Division in repulsing the sortie made by the enemy from that fortress.

About the 14th March, having received clothing and necessaries, the 1st Battalion proceeded to join the division with Lord Wellington, and on its way to the army had the good fortune to save a division of heavy artillery that was on its way from Orthes, and which was threatened by a body of the enemy, who were detached by Marshal Soult round the right flank of Wellington's army, it was thought, for the purpose of destroying that convoy of ordnance. The battalion joined its division, the 4th, at the latter end of the month.

April 6th.—Crossed the Garonne, and encamped near Toulouse.

TOULOUSE.

April 10th.—At eight in the morning a general attack was made on the position of Marshal Soult. The result was that, after a day spent in sanguinary conflicts at various points, at night the allied troops were established on three sides of Toulouse, and in possession of the enemy's fortified position.

Lord Wellington was making arrangements for another attack on the night of the 11th; but the French retired, leaving three generals and 1600 officers and men prisoners. In this battle the 27th had the honour of leading the column of attack, and afterwards formed the left of the line, where many opportunities were afforded them of displaying the same firmness and intrepidity for which the Enniskillen regiment has ever been distinguished, being exposed during the entire day to a galling fire of artillery and musketry, and threatened by a body of cavalry on the flank. The loss of the regiment on this day was two officers (Captain Bryant and Lieutenant Gough) killed, five officers (Colonel Sir John Maclean, Captain Geddes, Lieutenants Hamett and Byrne, and Ensign Armitt) wounded. There were two sergeants and 21 rank and file killed; two sergeants, one drummer, 73 rank and file wounded. On the 10th the battalion followed the flying enemy to Villefranche, where intelligence was received that Napoleon had abdicated the French throne and that Louis XVIII. was proclaimed. The battalion marched at the latter end of April to Bordeaux, and encamped near that city.

In June it embarked for North America, arrived at Quebec in July, and landed at the Three Rivers. The services of the 3rd Battalion in America are detailed with those of the 1st Battalion, both battalions having acted together in the advance to and retreat from Platsburgh. The 27th arrived at Portsmouth from America on the 3rd of August, and immediately proceeded to Ostend, and thence to Paris.

3RD AND 2ND BATTALIONS DISEMBODIED.

After having given drafts to the 1st Battalion at Paris, the 3rd Battalion returned to England in February, and was disembodied in that month at Canterbury. It was embodied ten years and six months, seven years and six months of which it was actively employed in Spain, Portugal, France, and America, and no man could have more honourably upheld the high character of their corps for gallantry and discipline than this battalion of young Enniskilleners on every occasion.

The 2nd Battalion was disembodied at Dublin in April, 1817, on being joined in that garrison by the 1st Battalion from France. This battalion was embodied at Glasgow in November, 1804, and on the following year it was embarked for foreign service, on which it was employed until the peace. The gallant services of this battalion are detailed with those of the 1st, having acted for nearly the whole period of its existence on the same service with the 1st Battalion.

HONOURS OF THE 27TH.

In commemoration of the services of the 27th Regiment in this war, the following honorary distinctions are borne on their colours and appointments:—

Badajoz, Salamanca, Vittoria, Pyrenees, Nivelle, Orthes, Toulouse, Peninsula, Waterloo.

ENNISKILLEN AND ITS REGIMENT.

The following interesting souvenirs of our local regiment —1st and 2nd Battalions—I cull from the *Erne Packet* of May 22nd, 1817. The peace of Europe had become an established fact at that time, nearly two years after the defeat of Napoleon on the field of Waterloo; and it being no longer deemed necessary to keep up a second battalion, it will be seen that the corps was about to undergo immediate disembodiment, after departing from its native town, where it had been ten months stationed, and had won during that time the respect and esteem of all classes of the inhabitants. The poorer classes had felt much misery from the high rates that prevailed for articles of daily subsistence. The brave fellows comprising the 2nd Battalion, as well as their brethren of the 1st, were not unmindful of their fellow-townsmen, and nobly combined with devotion to their sovereign a feeling of Christian charity to those who needed such a manifestation of sympathy at home. The 1st Battalion had not yet returned from France; but its subscription of upwards of £70 showed unmistakably the generous impulses which swayed every heart belonging to it,—a most gratifying evidence that zeal for one's sovereign, even on a foreign field, does not obliterate the finer feelings of our nature. I may remark that Colonel G. J. Reeves, of the Inniskillings, had just been voted the freedom of the city of Dublin, for the distinguished part he had taken in the war. This gallant officer had been wounded at Catalonia. But, to proceed:—

"At a numerous meeting of the inhabitants of the town and neighbourhood of Enniskillen, convened at the Town-hall, on Tuesday, May 8, 1817, by public notice,

"The Rev. Thomas Johnston, Provost, in the chair,

"It was unanimously resolved, on the motion of John Armstrong, Esq., seconded by the Rev. Dr. Burrowes,

"That the following Address, expressive of our high opinion of the conduct of the 2nd Battalion of the Enniskillen Regiment while quartered in our town, and of our regret at their being removed from this garrison, be transmitted by the provost to Colonel Sir John Maclean, with a request that he will communicate the same to the officers, non-commissioned officers, and privates of the battalion.

"Resolved—that the address, with the answer of Sir John Maclean, be printed in the *Erne Packet* and *Correspondent* newspapers.

"THOMAS JOHNSTON, Chairman.

"SIR,

"We, the inhabitants of the town and neighbourhood of Enniskillen, beg leave to express to you, and to the excellent battalion under your command, our sentiments of sincere regret at your removal from your quarters in this district. We had, through many arduous campaigns, followed with our anxious wishes the victorious progress of our brethren in arms, and, at the report of general peace, looked on it as a high gratification to be honoured by the enjoyment of social intercourse with any of those warriors who had successfully supported the character of their country abroad. But in the presence of the regiment which traces its first origin to our town, and fills its ranks from this country, we had a concern of deeper interest; and in proportion to the warmth of family affection which was experienced at having a battalion of the Enniskilleners quartered among us, must be our feeling of sorrow at your departure. For we can with truth, sir, assure you that the pride with which we had received the accounts of your various successes in the field was exceeded by the pleasure which a residence of ten months among us afforded, in the opportunity of witnessing the loyal and active spirit, and the disciplined propriety of conduct, which distinguished the 2nd Battalion of the Enniskillen Regiment.

"In the amicable intercourse of the battalion with the inhabitants we have had a satisfactory proof that those sentiments of affectionate regard were reciprocal. And we acknowledge with gratitude the ample testimony of charitable kindness the battalion gave, in its munificent donation to our fund for the indigent, and their kind recommendation of our charitable institution to the battalion of the regiment then serving abroad. The liberal contributions of the whole regiment of Enniskilleners, which exhibit so honourable a union of the compassion of the brother with the manly liberality of the soldier, call forth our thanks.

"To you, sir, in particular, we present them, with earnest wishes for your health and prosperity; and we request that you will please to make known to the officers, non-commissioned officers, and privates of the 2nd Battalion, the grateful and affectionate feelings of our hearts. And, though we may regret that we shall not continue to have our friends of this battalion embodied in one regiment, to exert their efforts collectively in the cause of their king and country, we have the fullest confidence that, on their return to cultivate the arts of peace in this and other parts of the empire, we shall find the Enniskilleners exercising, individually, as useful and loyal citizens, virtues not inferior to those which they have, with general applause, displayed as soldiers.

(Signed) "THOMAS JOHNSTON, Chairman.

"To Colonel Sir John Maclean, K.C.B.,
2nd Battalion 27th Regiment.

[REPLY.]

"Dublin, May 13th, 1817.

"MY DEAR SIR,

"I yesterday received, with much pleasure, the highly flattering Address of the inhabitants of the town and neighbourhood of Enniskillen, on the departure from hence of the 2nd Battalion of the 27th Regiment; and I have not failed to communicate the same to the officers and men under my command. Believe me, sir, the regret at parting with the battalion, so handsomely expressed in the Address, is perfectly reciprocal.

"Our sovereign having no further occasion for our services at present, we are about to retire to our respective homes; and every individual will, I am sure, carry along with him a deep and lasting impression of the kindness we met with at Enniskillen. And we shall always recollect, with pride and satisfaction, the happy period that gained us the good opinion of its inestimable inhabitants.

"I have the honour to be, my dear sir,

"Your faithful and obedient servant,

"JOHN MACLEAN, Colonel.

"To the Rev. Thomas Johnston,
&c., &c., &c.

"The subscription of the 1st Battalion, 27th Regiment, forwarded from France for the relief of the Enniskillen poor, was accompanied by the annexed letter:—

"Doignies, near Cambrai, 15th February, 1817.

"SIR,

"The present distressed situation of the poor and labouring classes of the people in Ireland having been communicated to the 1st Battalion of the 27th or Enniskillen Regiment, the officers, non-commissioned officers, and privates have unanimously agreed to subscribe a day's pay towards relieving the poor of the native town of the regiment. In consequence of this resolution, I have much pleasure in transmitting to you an order for £70 19*s.* 10¼*d.*, being the amount of the subscription; and am, sir, to request that you will be pleased to dispose of it in such manner as may appear to you best calculated to give effect to the wish and intention of the subscribers, namely, the relief of the poor distressed inhabitants of Enniskillen.

"I have the honour to be, sir,

"Your obedient servant,

"C. THOMPSON,

"Lieutenant-Colonel commanding 1st Battalion 27th or Enniskillen Regiment.

"To the Worshipful the Provost of Enniskillen.

"A meeting of the Benevolent Society was held, on receipt of the above, at which the following reply was unanimously voted, and transmitted by the provost to the commanding officer, after the arrival of the regiment in Dublin:—

"The Benevolent Society of Enniskillen discharge a most agreeable duty in returning their cordial thanks to the commanding officer of the 1st Battalion, 27th Regiment, and through him to the officers, non-commissioned officers, and privates composing that corps, for their munificent donation in aid of the funds of the institution for the relief of the poor of the town and neighbourhood.

"In duly estimating the generous sympathy evinced on the occasion towards the distressed poor of the native town of the 27th Regiment, the enhancing circumstance that it emanated from the battalion when serving their country in a distant land should not pass unnoticed, as practically blending social and Christian obligation with the most arduous military duties. More than twenty years of absence, of perils, and of glory had in no degree weakened the ties of affinity, or the recollections of home; and a meed more sacred than even the civic wreath now adds its lustre to the hard-won laurels, with which successive campaigns have covered the Enniskillen Regiment.

"The Society beg to accompany the present warm expression of their gratitude and regard with an assurance that their anxious attentions have been, and will continue to be, exerted to apply the donation in the manner best calculated to alleviate the sufferings of the poor of Enniskillen, and thereby give effect to the benevolent intentions of the regiment.

"To the Commanding Officer of the
1st Battalion of the 27th Regiment."

As a result of the peace, the Army Estimates showed a reduction of 55,343 men, at a saving of £1,800,000 for the year.

FOREIGN SERVICE.

1817.—In May the regiment marched into Dublin, after an absence on foreign service of twenty-four years. On July 18th, 1818, the regiment marched to Waterford, and in December embarked at Cork for Gibraltar. The years 1819—1823 were spent in garrison at Gibraltar. On December 1st, 1823, embarked at Gibraltar for the West Indies, arrived at Barbadoes on January 4th, 1824, and sailed again on the 7th to be quartered at Demerara and Berbice.

1824–1826.—In garrison at Demerara and Berbice. During the three years that the regiment remained in these colonies it suffered most severely, having lost in the above period 241 persons. The years 1826—1830 were spent in St. Vincent, Grenada, and Barbadoes.

HIGHLY COMPLIMENTARY.

The following is a copy of the farewell order issued to the 27th Regiment by his Excellency Sir James Lyon, lieutenant-general commanding the forces in the West Indies, on their return home :—

"Head-quarters, Barbadoes, 22nd November.

"The 27th Regiment being on the eve of embarkation, the Lieutenant-General commanding, in the separation from so valuable a part of his force, begs to convey to Lieutenant-Colonel Hare, his officers, non-commissioned officers, and men, his anxious wishes for their prosperous voyage, happy landing, and future success.

"Sir James Lyon must ever bear in recollection the zeal with which the 27th, whilst serving with him, have performed every duty, and he views, therefore, this departure with sincere regret; but his knowledge of their former more active and splendid services satisfies him that, to whatever destination the commands of their sovereign may hereafter direct the Enniskillen Regiment, they will maintain that distinguished reputation which has secured to them the respect and applause of those under whom they have served.

"By order of the Lieutenant-General commanding,

"H. Craig, Lieutenant-Colonel D.A."

AT HOME AGAIN.

1831.—The regiment landed in 1831 at Cork, and were quartered in Fermoy. They were afterwards removed to Limerick, and during the next year were quartered at Fermoy, Buttevant, Limerick, Ballinrobe, with detachments in various towns in Limerick, Roscommon, and Mayo. On the 18th June the company quartered at Castlebar moved to Ballyshannon, 1st do. from Ballinrobe to Donegal and Pettigo, 1st do. from Ballinrobe to Cavan. The detachment from Newport moved to Castlebar, 1st do. from Castlebar to Cavan; and the detachments in Westport and Tuam moved on 21st June to Enniskillen. On the day following the head-quarters marched from Enniskillen, which afterwards received the addition of several companies out on detachment. The regiment was spent during the first half of 1834 in quarters, detachments being in Omagh and various surrounding towns. On the 4th July head-quarters moved to Mullingar. In 1835 the regiment proceeded to Cape of Good Hope, where they were busily employed till 1848.

The following is a list of officers who embarked with the service companies:—Lieutenant-Colonel Hare; Major M'Pherson; Captains Johnstone, Amsinck, Sleator, Stanford, Maclean, and Fawkes; Lieutenants Durnford, Williamson, Butler, and Cholmely; Ensigns Foster, Watson, Fignolles, Lewes, King, and Hare; Paymaster Raymond; Adjutant Edden; Quartermaster Thompson; Surgeon Mostyn.

CAPE OF GOOD HOPE.

The following is a digest of the services of the 27th Regiment from 1st January, 1835, till 31st December, 1840.

The regiment disembarked at Algoa Bay on the 2nd September. On the arrival of the regiment at Graham's Town on the 8th of September, a provisional battalion, consisting of 400 Hottentots, divided into four companies and officered by colonists, was attached to the regiment, the 72nd and 75th Regiments having similar battalions attached to them.

At this period the eastern frontier was overrun and desolated by the Kaffirs, a barbarous nation on the confines of the colony; and in consequence of the alarming state of the country, the British settlers (several having fallen victims to the murderous disposition of the savages) had abandoned their homes and property, and taken refuge in Graham's Town. The country continued in this state of confusion and alarm, but the Kaffirs were finally overcome, and their country conquered and occupied by British troops.

On the 22nd October following, a treaty of peace having been concluded with them by Sir Benjamin D. Urban, then governor and commander-in-chief at the Cape of Good Hope, the head-quarters and two flank companies marched for Cape Town, leaving the major and four companies of the regiment, with the provisional battalion attached, on the frontier. These remained on the frontier service until the beginning of the year 1837. About that time the conquered province called "Queen Adelaide" having been, by orders from home, delivered over to the Kaffirs, and abandoned by the British troops, the provisional force of Hottentots was disbanded. These men continued faithful to the British Government during the Kaffir war. They were willing and excellent soldiers, particularly for patrol duty.

The four companies marched on the 17th March, 1837, and joined head-quarters at Cape Town on 12th April, 1837, on which occasion the following order was issued by Lieutenant-Colonel Sir Benjamin D. Urban, commander-in-chief at the Cape of Good Hope:—

"The Commander-in-Chief avails himself with great pleasure of the occasion presented by the arrival of the four companies of the 27th Regiment, so long detached under the command of Major M'Pherson, to publish the following extract from a frontier order of the Commandant of the district in which they have been serving—an order highly creditable to Major M'Pherson and the officers, non-commissioned officers, and soldiers of those companies; and he requests at the same time to congratulate Colonel Hare upon the excellent regimental discipline, which has stood the test of the trying circumstances in which those companies have been placed, and so honourably carried themselves through it:—

"'The Commandant cannot take leave of the 27th Regiment without his expressing to Major M'Pherson the high sense he entertains of its efficient services on the frontier. From the period of the arrival of the regiment on the frontier it was necessarily greatly detached, and the command of very important posts devolved upon several officers of the regiment. These commands have been conducted by those officers fully to the satisfaction of the Commandant, and to the advantage of the public service. The Commandant desires further to express to Major M'Pherson his warmest thanks for the great support and assistance he has afforded him upon all occasions, and to request Major M'Pherson will do him the favour to express to the officers and men of the regiment his approbation of the general character of the regiment, and which he shall have great pleasure in bringing under the notice of his Excellency the Commander-in-Chief of the Forces.'"

The regiment remained doing garrison duty in Cape Town until July, 1838, when Colonel Hare was appointed acting lieutenant-governor of the eastern province of the Cape of Good Hope; and having proceeded to assume the government, the regiment was ordered to follow him to the frontier, and embarked at Cape Town in the month of October, 1838, arriving at Graham's Town in the same month. At the close of 1839 Colonel Hare, having been confirmed in the civil government, resigned the command of the regiment, which he had held with distinguished credit since the year 1825. The command then devolved upon Lieutenant-Colonel M'Pherson, who again joined the service companies, the command of the regiment being in the interim held by Major Johnstone.

THE DEPÔT—1836.

After the embarkation of the service companies in May, 1835, for the Cape of Good Hope, under Colonel Hare, the four depôt companies, being left under the command of Major Maclean, proceeded on the 25th of the same month from Dublin to Newbridge, and thence on 24th June following to Nenagh, where it remained until August, 1836, when it was ordered to the English establishment, and embarked at Cove for Chatham on the 30th August, 1836. Officers remaining with depôt: Major Maclean; Captains Manly, Goodman, Byrne, and Smith; Lieutenants Lonsdale, Vereker, Roberts, W. Butler, Tounzel, Sparke, and M'Call. At Chatham the depôt remained doing garrison duty to the satisfaction of Lord Hill, the commander-in-chief, and the adjutant-general, as also Sir William Wane, the commandant of the garrison. In July, 1838, Major M'Pherson arrived in England from the service companies and assumed the command of the depôt, at which time (23rd July, 1838) they removed to Dover, and continued there until 9th September, 1839.

FESTIVAL TO DUKE OF WELLINGTON.

During their stay at Dover this depôt was much noticed by his Grace the Duke of Wellington, then Lord Warden of the Cinque Ports, and other eminent and distinguished characters, and their general conduct was so good, that at the memorable festival given by the Cinque Ports to his Grace in the month of August, 1839, the committee appointed to manage the festival made special application to Lord Hill, the commander-in-chief, requesting that the depôt of the 27th Regiment might remain in Dover until after the entertainment was over, it being their desire that the Inniskilling regiment should furnish the guard of honour on the occasion.

The request was complied with, the commanding officer and all the officers being present at the entertainment.

In the month of October, 1839, the depôt was marched from Dover to Gravesend, and embarked for Ireland. On arriving at Cork, they received orders to proceed to Waterford, and remained there during the winter.

NEW COLOURS.

On 24th May, 1837, the regiment paraded at twelve o'clock, when new colours were presented to the regiment by Lady Catherine Bell. The notices of the years 1840–1844 are occupied with accounts of marchings and detachments, and reliefs.

The following were the officers in 1841:—Lieutenant-Colonel M'Pherson; Major Goodman; Captains Sleator, Maclean, Byrne, Smith, and Lonsdale; Lieutenants W. Butler, Neynoe, Smith, Lewes, and Irwin; Ensigns Stapylton, Molesworth, and Tunnard; Adjutant Midgley; Quartermaster Thompson; and Paymaster Raymond.

NATAL.

The year 1842, by far the most eventful year in the annals of Natal, was ushered in by the appointment of the officer commanding the Umgazi post, Captain T. C. Smith, of the 27th Regiment, as commandant of Natal.* The force despatched on this expedition consisted of 100 men of the 27th Regiment, which left on the 25th January with orders to proceed to the Umgazi post, thence to march with 200 men upon Natal, leaving 50 men at the camp to act as a post of observation, and to maintain the overland communication with the colony. Two field-pieces only were sent with the detachment. The inadequacy of such a force, too large for negotiation and too small for conquest, against a body of unerring marksmen, safe in position, was at once exposed by the press and by every intelligent person in the colony.

An account of the march to Natal—its extreme difficulties and dangers—its exhausted state on its arrival there on the 4th of May, are graphically described in a letter of an eye-witness, one of the troops :—

"12th May, 1842.

". . . . I received your epistle on the south bank of the Umsacoola river. It was after travelling 18 miles, and the whole day raining on us, and then had to mount guard the same night. . . . .

"The three buglers were divided every day into three divisions, one with the artillery, and one with the division, and one with the rear-guard. Our march chiefly was along the south-east coast the whole way, until within a few days of Natal. . . . . The men caught three brown bucks and gave them to the officers; we saw a great many sea cows, and came across the spoor of lions and elephants in the woody parts along the coast. We *suffered much from marching in the sand; it got into our boots and cut our feet to pieces, and*

* From "Natal," Part I., by J. C. Chase, kindly lent me by Lieutenant-General Smith, who also gave me other valuable assistance.

the sun reflecting the sand burned our faces. In like manner the men had many fatigues in repairing the wads every four or five miles they went along. Mr. Archbell, the Wesleyan missionary, and family, were in company with us the whole way. . . . .

"We crossed 122 rivers, and the most of them *we had to swim over;* some of them extending 600 and 700 yards in breadth: they are the largest and greatest rivers I ever saw in my life. We stopped two days and part of the third at the Umkomass river. During our stay here we had muster-parade and articles of war. The night before James Dentt, of No. 2 Company, died, and was buried next evening with the usual martial ceremonies. Poor fellow! his death was occasioned from *the fatigues of the march,* and it is a wonderful mercy of Providence that a great many more did not share the same fate. The next day we departed from the river; and when the guns went over they loaded with grape, and every company, according as they reached the other side, all loaded with ball, and every soldier in the expedition, from the captain, did not know the moment the enemy might approach him, and due precaution was taken every night in pitching the camp. The guns and infantry were ordered to be kept loaded until further orders. The next day's march we came within 25 miles of Natal, and that evening there came four Englishmen out to meet us, all armed with swords, pistols, and double-barrelled guns. . . . .

"But to draw to Natal. The last day's march being the 4th May, as we drew within 12 miles of the town, it was surely handsome to see all the pretty cottages and handsome villages belonging to the peaceable Dutch farmers. The captain received word outside of the town about the enemy abandoning the town and fort. We arrived in Natal about four in the evening, and pitched our camp on a projecting hill, about the distance of six miles from the town and harbour; for the captain thought the enemy might give him a visit that night, and all preparation was accordingly made in placing the guns and waggons all round the camp. . . . . The English agent *wondered very much to see such a small force* going to face the enemy as we were, and he asked the captain if there was not a force coming by sea; but the captain told him there was not, and that he was not the least afraid to meet as many more. The agent smiled to think he would face 1500 men armed in the manner they were, with swords, pistols, and double-barrelled guns—the *best armed men in the colony*. . . . The evening we came in here we saw the haughty Dutch banner was displayed on the fort at the harbour, as large as life. But the next morning the captain and the engineer officer, with all the Cape corps, and a few of the artillery, went down to the fort and *hauled down the rebellious flag, and hoisted the British Union* of old England, and spiked their gun alongside of it, a 6-pounder. In the mean time the captain and engineer officer planned out a place for our camp, alongside of the town, but in an open plain. The captain and the remainder of them arrived at the camp about four o'clock in the evening. . . . .

"The next day we marched to our camp, where we were destined to remain with fixed bayonets, and the officers with their swords drawn, and in full uniform. We passed through a small village belonging to the Dutch, called Kongela, but there seemed to be very few inhabitants in it, as they were all out in the country. We marched through the town and came to our camp ground at 11 o'clock; but such a place for *bad* water I never saw in my life: it is as *black as ink*, and full of different insects, and stinks into the bargain. I am very much afraid it will make away with the whole of us before long. . . . .

"The duty is very hard here. The men have only two nights in bed. . . . . I forgot to mention that the Boers bought up all the provisions in the town before we came up.

"To draw to a close with my narrative, I wish to mention that the Boers were not far astray of us being starved, for our *provisions are out these four days past*, and *we are living on handfuls of rice*; but thanks be to God, a small brig arrived in harbour this day with plenty of provisions. . . . . She brought two long 18-pounders with her, for to put upon our battery. . . . .

". . . . I can assure you that I and many of our expedition have not closed our eyes since we came here, and this is the ninth day since our arrival, and I am just the same as usual.

"John Brown,

"Bugler, 27th Regiment."

Things remained in this state for some days. The farmers collected in a force of about 300 men; and this, had Captain Smith not been bound down to suffer anything rather than proceed to extremities, he might easily have crushed in its rise—for it was fifteen days in collecting—but being tied down by his instructions, he was obliged to submit to observe an enemy raise a force before his eyes. This forbearance was construed by the Boers into fear, and this idea, added to the evil influence of the Dutch captain, Reus, brought matters to a most painful issue.

CONGELLA.

On Monday, the 23rd, the first aggressive act was committed. The Boers seized about sixty oxen, and then moved down upon the camp. Captain Smith then opened fire upon them with one 18-pounder, which he had just got mounted, and had not been in its carriage more than three hours. Believing it necessary to prevent a repetition of such an outrage, Captain Smith determined to attack the Boers' camp at Congella, three miles off, that very night. The force that accompanied Captain Smith consisted of one subaltern and 17 privates Royal Artillery; 1 subaltern, 1 sergeant, and 7 privates Royal Sappers; 2 captains, 2 subalterns, 5 sergeants, and 100 rank and file 27th Regiment; and 2 mounted orderlies of the Cape Rifles.

The party arrived without molestation within 800 yards of the place, and a boat, with a howitzer, not having arrived at the place as intended, the attack had to be made without its valuable assistance. Captain Smith having given the order to advance, the troops had just moved to where a range of mangrove bush opened to a level space in front of the Congella, when a heavy and well-directed fire from the bush was poured upon them; upon which they immediately formed and commenced a fire in return, while the two 6-pounders were loading. A destructive fire from the guns silenced the enemy for a while, but several of the oxen becoming wounded, caused great confusion in the ranks, and great delay in reloading. The Boers again opened a heavy fire, and a severe loss resulted in consequence; the order to retire was given, the death of the oxen rendering it impossible to remove the guns. The Boers followed up their partial success by attacking Captain Smith in camp, but they were obliged to retire about an hour before daybreak with severe loss.

Lieutenant Wyatt, of the Royal Artillery, was killed early in the action, and Captain Lonsdale and Lieutenant Tunnard were severely wounded. The total number of the killed was 1 subaltern, 15 rank and file; wounded—1 captain, 1 subaltern, 2 sergeants, 27 rank and file; missing—1 rank and file.*

A TRYING SIEGE.

On the following day, the 26th, the north-eastern point of the bay, on which the provisions and an 18-pounder, landed from the *Pilot*, had been stored, fell into the hands of the Boers. Captain Smith was thus cut off from his supplies. The men were placed upon half allowance, and their position secured as well as possible. On the morning of the 31st the farmers made a desperate attack on the camp, throwing into it during the course of the day 122 round shot, besides keeping up an incessant fire of musketry. On the following day 124 round shot were thrown in; and on the 2nd June opened a fire from the 18-pounder, which they had contrived to bring from the Point, while they still continued their discharge of musketry.

Finding that the few cattle remaining at the Kraal were dying, either from wounds or want of sustenance, Captain

* Captain Smith's despatch to his Excellency Colonel Hare, C.B., K.H.

Smith directed that they should be killed and made into *biltong*, reducing the issue to half a pound daily. The practice from the camp was excellent, a shot from the 18-pounder having dismounted one of the enemy's 6-pounders on the 3rd June, besides wounding several of those attached to it. On the night of the 6th a party destroyed some of the enemy's works without any loss. In a subsequent sortie on the night of the 18th the Boers were surprised in the trenches, and bayoneted after a stout resistance. In this attack, which was headed by Lieutenant Molesworth, 27th Regiment, Ensign Prior and two privates were killed, and four others severely wounded.

At this period only three days' issue of meat remained. The remaining horses were thereupon killed and made into *biltong*. Captain Smith had hitherto been issuing biscuit dust, alternating with biscuit and rice, at half allowance. The horseflesh was used for the first time as food on the 22nd inst.

On the 11th of June a detachment of the 27th Regiment, 100 strong, had been embarked from Port Elizabeth for the relief of Captain Smith, and a considerable force was also despatched from Cape Town; and on the 26th Colonel Cloete landed, with detachments of 25th and 27th Regiments, and relieved the post in gallant style between three and four o'clock in the afternoon. In his despatch to Colonel Hare, Captain Smith reported—"Nothing could exceed the patience and cheerfulness evinced by the troops under the privations they suffered, and I feel confident that, had it been necessary to have held out for a longer period, they would have endured their further continuance without a murmur."

In his despatch to Sir George Napier, K.C.B., governor and commander-in-chief of the Cape, he thus wrote:—

"The gallantry with which Captain Smith defended his post for a whole month, under no ordinary circumstances of privation, having been reduced to horseflesh for food, closely hemmed in by a desperate and vigilant foe, with no less than twenty-six wounded within his closely confined camp, is highly creditable to him and his party. Thus was accomplished within the incredibly short space of one month from the date of Captain Smith's report of his position the relief of his party, at a distance of 1500 miles from Cape Town, whence the relief was dispatched, his communication having had to pass through hostile bands and a savage country."

To mark the sense of the services of Captain Smith for "maintaining his post at Port Natal against an overwhelming force of rebels, from the 24th of May to the 26th of June," a large subscription was immediately set on foot to purchase him a sword, in which both the friend and enemies to the late pacification joined. For his gallantry in this action Captain Smith was promoted to the rank of major. On the sword-blade was the following inscription:—

"Presented to Major T. C. Smith, H.M.'s 27th Regiment, Inniskillings, by British subjects of the Cape of Good Hope, to express their sense of his indomitable bravery in maintaining his post at Port Natal against an overwhelming force, from the 24th May to the 26th June, 1842."

HOME SERVICE.

The head-quarters remained in Graham's Town until the 1st February, 1846, on which day a wing of the regiment, under the command of Captain Durnford, was ordered back to the frontier; and on the 27th March the lieutenant-colonel and the remainder of the corps followed, and were employed on the frontier until January, 1848, when the regiment was ordered to Cape Town, for embarkation for home service. Sailed from Table Bay on 12th February, and disembarked at Gravesend on 19th April. The regiment was moved into Chatham garrison, where 76 of the old soldiers were discharged. The remainder, composing the head-quarters, proceeded to Edinburgh, thence to Glasgow (where the depôt was quartered), arriving in that garrison on the 26th April, 1848.

SCOTLAND.

The regiment remained in the garrison of Glasgow for two years, with eight companies at head-quarters, furnishing a field officer's detachment at Paisley and an officer and twenty men to Dumbarton Castle.

When the service and depôt companies of the 27th Regiment were consolidated at Glasgow, Major-General Henry J. Riddell deemed it his duty to make a special report to the adjutant-general of the great satisfaction he had received from the remarkably orderly and soldier-like conduct of the depôt companies since they had been stationed in North Britain; and he requested, by his letter of 26th April, 1848, the adjutant-general to bring it under the notice of his grace the commander-in-chief, who would observe from the accompanying report, that of a body of 600 men there were only fifteen sick, two absent without leave, and not a prisoner either in the district military prison, the garrison cells, or in the custody of the civil power.

In reply thereto, Adjutant-General John Macdonald, by letter dated from the Horse Guards, 28th April, 1848, stated he had submitted to the commander-in-chief General Riddell's letter, with inclosure, and that he was directed to acquaint him that his grace had derived the greatest satisfaction from the perusal of those documents, the showing of which was so creditable to Brevet-Lieutenant-Colonel Goodman and the depôt companies of the 27th Regiment.

In April, 1849, the regiment was called upon to furnish detachments to Fort William and Fort Augustus. On the 5th April, 1850, the regiment was moved from Glasgow to Stirling Castle; the detachment for Paisley, Dumbarton Castle, and Fort Augustus rejoining head-quarters. The regiment remained stationed here until November, 1850, when it was ordered to Ireland. The six companies on detachment embarked at Glasgow on the 12th November for Belfast, and the four head-quarters' companies on the 14th, arriving at Belfast on the following morning, and occupying the barracks there.

On the regiment leaving Scotland the following letter was addressed to Lieutenant-Colonel Magennis, commanding the regiment, by Major-General Riddell, commanding the troops in Scotland:—

"Adjutant-General's Office,
"Edinburgh, 13th November, 1850.

"SIR,

"I am directed by the Major-General commanding to inform you that he cannot allow the 27th to embark without expressing the great satisfaction he has derived from the exemplary conduct of the regiment and depôt while serving in North Britain. And I am at the same time to assure you that it is a source of great regret to the Major-General to lose the services of so distinguished and efficient a corps.—

"I have, &c.,

"JOHN EDEN,
"Assistant Adjutant-General."

1852.—On the 14th April Lieutenant-Colonel Cunynghame joined to take command of the 27th Inniskillings, in succession to Lieutenant-Colonel Magennis, who was appointed to the situation of Inspecting Field Officer at York.

NEW COLOURS.

On the 24th day of May, 1852, in Dublin, the anniversary of the taking of St. Lucia in 1796, the 27th Inniskillings were paraded under Lieutenant-Colonel Cunynghame, to receive from his hands their new standard of colours. At 2.30 p.m. the regiment was formed into square, the new colours in charge of the junior ensign, the old colours in charge of that of the senior. The lieutenant-colonel having called the regiment to attention proceeded to address them. After recapitulating at great length the whole of the services of the corps from its earliest formation to the present time, dwelling more particularly upon the reminiscences of the 24th May, 1796, upon which it so particularly distinguished itself at the capture of the fort of

Morne Fortuné, in the island of St. Lucia, he concluded his harangue in the following terms: "And now that I have run through the services of the regiment from its earliest formation until the present moment, I will but add a few words explanatory of my own feelings in regard to the emblems which it has devolved upon me to place into the keeping of the corps. It is her Gracious Majesty's pleasure that we receive these glorious banners, and from her also emanate rules and ordinances for our general conduct and guidance. We must therefore always bear in mind that if we neglect or disobey the latter we cast a slur upon the former, and that as each soldier of the Inniskillings looks with pride and pleasure upon the standards of his corps, so must be his constant regard for its character and unsullied name, looking with pride and satisfaction to that day when he first became enrolled under its banners."

The lieutenant-colonel then presented the new colours to the senior ensign, the junior receiving the old ones. A double line was then formed, facing inwards, a general salute was given in honour of the old colours, and the same for the new, and the regiment then retired to their quarters.

FUNERAL OF DUKE OF WELLINGTON.

On the 15th November a deputation, consisting of one field officer (Lieutenant-Colonel Cunynghame), one captain (Captain Tounzel), one subaltern (Lieutenant Manly), one sergeant, one corporal, and six privates, were ordered to London to attend the funeral of Field Marshal his Grace the Duke of Wellington, the late commander-in-chief, to represent the 27th Inniskillings. This deputation returned from London on the 23rd of the same month, and the soldier-like appearance and good conduct of the party met with the approbation of the authorities of the Horse Guards, when inspected at Chelsea by the general commanding-in-chief.

ENNISKILLEN.

On the 18th March the regiment, under the command of Lieutenant-Colonel Cunynghame, marched out of Dublin, its head-quarters, to be stationed at Enniskillen, with detachments at the following places:—Armagh, Sligo, Cavan, Ballyshannon, Belleek, Charlemont, and Downpatrick. The 27th Inniskillings had been stationed in Dublin for the space of twenty months, during which period their conduct and general behaviour had elicited the constant praise of every authority, both civil and military. They were received at Enniskillen with immense enthusiasm, bells pealing, &c., a welcome home to their native city after their toilsome services and long absence.

[*Regimental Order.*]

"Enniskillen, 13th March, 1853.

"The following letter having been received from Lieutenant-General Sir W. Napier, K.C.B., on his taking leave of the 27th Regiment, is published for the information of the corps:—

"Scinde House, Clapham, 17th October, 1853.

"Sir,

"Having this day received official notification that her Majesty has been graciously pleased to appoint me to the command of the 22nd Regiment of Foot, I have the painful task of taking leave of the 27th Regiment.

"For personal objects that should never have been the case: so proud was I of obtaining the 27th Regiment, that when the late Duke of Wellington offered me the 43rd Regiment, with which I had always served, and with which my military recollections and sympathies were bound up, I refused it—with painful feelings, I acknowledge, but with a full conviction that to relinquish on personal grounds the honour of commanding the 27th Regiment would be improper. You will therefore believe that on no light grounds I do so now, and I am sure that the regiment will feel that an honour to mark the Sovereign's approbation and esteem of my late brother's services could not be declined. The honour was to him, not to me; and I therefore quit the regiment, to which I feel attached because it is, and always has been, *unsurpassed* in glorious action, and in good conduct *unsurpassable.*

"I remain, sir, with every warm wish for the prosperity of the regiment, generally and individually, your obedient servant,

(Signed) "W. NAPIER,

"Lieutenant-General.

"The Officer commanding 27th Inniskilling Regiment."

On the 14th December, 1853, Lieutenant-Colonel Cunynghame assembled the Inniskillings, and informed them of his intention of relinquishing the command of the regiment. He expressed to them his deep regret that family circumstances compelled him to adopt a course which he felt most keenly. After explaining most fully the system of command which he had universally followed, a system which clearly demonstrated to him that strict impartiality invariably met with the respect of all, he concluded by trusting that on some future day he might yet have the gallant and noble-hearted Inniskillings under his orders, and that if it were his good fortune to command them in active service in the field, there was but one spot he need look for them. He had commanded three regiments, and served in some more, but for none did he bear a greater affection than for the 27th, and he should ever remember and be thankful for the support he had received from officers, non-commissioned officers, and the excellent and noble feelings which prevailed amongst the men. He bid them a hearty farewell, and God grant them health and happiness in the honourable, though arduous, duties of their military career.

DUBLIN.

The 27th Regiment remained in Enniskillen till March, 1854, when it received orders to move to Dublin. The head-quarters, under Major Williamson, marched out of Enniskillen on the 8th, and arrived in Dublin on the 11th March, 1854. All the detachments arrived in Dublin from their respective out-stations during the month, with the exception of those at Charlemont, Armagh, and Cavan, the two former arriving together on the 8th May, and the latter on 23rd May, 1854.

The *Fermanagh Reporter* of March 9th, 1854, thus referred to the regiment's departure:—" 27th REGIMENT. This fine body, which is peculiarly connected with home, took its departure yesterday for Dublin about half-past nine a.m. Lord Enniskillen and a number of the most influential inhabitants of the town presented an address in the barrack square to Major Williamson and the men under his command. The address was read by the Rev. Mr. Bradshaw, and responded to by the gallant major in a short but able and eloquent extempore address from the saddle. The men immediately marched out of the town, and were accompanied for miles by many of the inhabitants. They were our own people who were going, and the streets were thronged with people to look—perhaps their last—on a body of brave and well-behaved men, most of whom they will in all probability never look upon again." *

CORK.

The 27th was moved to Cork in June, 1854, to embark for foreign service. Half the corps, consisting of the left and part of the right wing, under the command of Major Tounzel, proceeded by railway from Dublin, arriving in Cork the same day (16th June). The head-quarters, consisting of the remainder of the regiment, under Lieutenant-Colonel Williamson, arrived in Cork the following day (17th June, 1854).

* It was then believed they were intended for the Crimea.

FOR INDIA.

On the 22nd June, 1854, the first division of the 27th Regiment, under the command of Major Tounzel, embarked from Queenstown on board the *Soubhadar*, for Calcutta. The first division consisted of Nos. 4, 6, and 8 Companies, and 9 men of the light company. Total, 8 officers, 13 sergeants, 2 drummers, 9 corporals, 179 privates, 16 women, and 27 children. Grand total, 254. The second division, under the command of Captain Stapylton, embarked on board the *Charlotte* on the 26th June, from Queenstown for Calcutta. The second division consisted of Nos. 5 and 7 Companies, 18 men of the grenadiers, and 11 of the light company. Total, 5 officers, 6 sergeants, 2 drummers, 9 corporals, 146 privates, 14 women, and 26 children. Grand total, 208. The third division, under the command of Major Durnford, embarked on board the *Barham*, from Queenstown, on the 4th July, 1854. Total, 8 officers, 12 sergeants, 3 drummers, 15 corporals, 214 privates, 31 women, and 50 children. Grand total, 333. The head-quarters division, under the command of Lieutenant-Colonel Williamson, embarked on board the *Southampton*, from Queenstown, on the 5th July, 1854. Total, 11 officers, 17 sergeants, 11 drummers, 10 corporals, 122 privates, 31 women, and 37 children. Grand total, 239.

Total strength of the corps embarked—3 field officers, 8 captains, 14 subalterns, 7 staff, 48 sergeants, 18 drummers, 43 corporals, 661 privates. On embarkation of the regiment for India, the depôt was left at Cork under the command of Captain W. D. Phillips. The grand total strength of the whole regiment was 1127.

A NIGHT OF HORRORS.

Perhaps there are few incidents in the career of the 27th Regiment about which less is known, than the details of that terrible night in 1854, when they suffered shipwreck, and many of them a cruel death, amidst the horrors of an "elemental war" in Algoa Bay. Band-Sergeant Presley, of the 27th, preserved the following account, which he has kindly given me :—

"On the evening of the 20th of September, 1854, a detachment of this old regiment was passing through an ordeal more terrific than that through which their comrades of other corps had passed on the same day over the blood-stained heights which bound the southern bank of the Alma.

"The only enemies that can terrify a British soldier are the elements and 'yellow-jack;' and it is not improbable that more men have been destroyed by shipwrecks, cholera, and other diseases, than by bullets of open foes, whom a soldier never fears. In the battle-field, in the midst of deadly conflict and the loud din of war, a soldier has a chance for his life; he is then in his natural element, and fully occupied in loading, firing, and preparing for the charge; but when threatened by devouring flames, or merciless waves, or both together, the case is altered. Yet under such trying circumstances the exalted discipline of the British soldier has always been conspicuous.

"There are but few regiments that have not at some period of their history been threatened with destruction by unnatural foes. Who that has not heard of the appalling position of the 31st Regiment in 1825, when the *Kent* East Indiaman, which was conveying them to India, took fire in the Bay of Biscay? The 81st and 82nd Regiments were shipwrecked more than once. In recent times the 43rd Regiment was wrecked in an African harbour which was full of sharks. The burning of the *Sarah Sands* (on board of which was a wing of the 54th) and the *Eastern Monarch* must be yet fresh in the memory of most readers.

"In the summer of 1854, the 27th Regiment embarked on board four transports bound for Calcutta. The transport *Charlotte* bore 227 souls. On the evening of the 20th September, 1854, they anchored in Port Elizabeth, Algoa Bay, for the purpose of replenishing their water casks. About eight p.m. a tremendous storm blew. In a few seconds both cables were sundered, and the vessel was at the mercy of wind and waves. None but the watch were permitted on deck; the remainder were in their hammocks partially undressed. The night was 'pitch dark,' and the surf rose mountains high. In three seconds after the cable was rent the ship struck against the reef. The shock was so violent that the occupants of the hammocks were thrown on the floor of the vessel. Two seconds afterwards she again struck and went to pieces. Then—

> 'Rose a cry from earth to heaven;
> Then shrieked the timid, and stood still the brave!'

"It is almost impossible to realize the horrible scene that ensued. By four a.m. not a single piece of the vessel could be seen, and 113 souls were dashed against the rocks; 114 were saved, or one more than half the number on board. Eleven women, twenty-eight children, eleven sailors, and sixty-three soldiers (27th) were lost. Only five women were saved. The survivors were treated with the greatest kindness by the inhabitants of Port Elizabeth, who crowded the shore and supplied the rescued with blankets and other comforts, and took them to their houses, where nothing was too good for them. For days afterwards, mangled bodies were found among the rocks, and so disfigured that some could not be identified. The corporation furnished coffins. In one grave ten bodies were interred. The inhabitants subscribed £400 for the sufferers, and supplied them with coloured clothes, in which uniform they sailed to Cape Town on board the man-of-war *Hydra*. After six weeks' stay, they embarked on board the *Maidstone* on the 9th of November, and landed at Calcutta on the 9th of January, 1855. Among the survivors still serving in the regiment are the two Sergeants (brothers) Fitzgibbon, Corporal Donelly, Sergeant Kingan, Privates Feeney, W. Elliott, James O'Hara, John Lawler, and McEnroe."

The following lines on the catastrophe were written at the time:—

> "A noble ship from Britain's favoured land,
> With many gallant hearts for India's strand,
> For water and shelter sought our bay,
> And proud she looked as she at anchor lay;

But how shall we this doleful story tell—
The sad disaster which her crew befel?
At five p.m. we saw her cable part,
E'en then we had misgivings in our hearts;
But as she bore upon the starboard tack,
And set her sail before the night was dark,
Our fears were lighter, and we hopeful were
That she might hold till morning should appear.
With straining eyes we watched her, till night
And distance hid her from our anxious sight;
Then hastening with our friends for social cheer,
We round our table sat without a fear.
At eight p.m. a signal gun was fired,
Then throwing down our game, the worst inferred,
And we sallied forth; the watchfire glares
Lightening the sky, awoke our fears.
Too true, alas! for when we reached the shore,
The hum of men mixed with the surge's roar
Burst on the ear,—we saw, with awful shocks,
The noble vessel strike upon the rocks.
All hope was gone!—for such an iron spot
Bespoke, too plainly, what must prove her lot—
Of all the places of our line of coast
She had selected one the very worst.
Twice was communication made by line,
And every effort made, but all in vain;
For twice it broke—nor could be thrown again.
At twelve, the lifeboat, with a gallant crew,
Pulled through the surf to try what she could do:
But as she neared the fatal ship, no rope
Was cast, and from their hold they broke.
Four times the gallant crew made the essay;
No rope was given, and then she could not stay.
Only one man—more lucky than his mates—
Leaped in the boat, was rescued from their fates.
The boat returning to the craggy shore,
Was stove amongst the rocks, and a wild uproar,
With generous aid the noble crew to save;
But Salmond nearly perished in the wave.
But now it seemed all means of help were gone,
And their sad fate was desp'rate and forlorn.
Six anxious hours we watched upon the strand,
To try to save, if any swim to land,
And ever and anon a wailing cry
Burst from the wreck, while waves all on her fly.
'A boat! a boat! for God's sake send,'
Was the wild cry we heard on waves and wind;
And as she lurched with each succeeding wave,
Her bending masts seemed pointing to their grave.

Meanwhile at every surf the vessel roll'd,
And direful shrieks another horror told;
Some hapless wretch, naked and numbed with cold,
Hopeless of life, despairing lost his hold.
And as each rolling wave succeeded wave,
Mother and children found a watery grave;
And oft, while some were floating towards the shore,
We heard their cry, but we could help no more.
At two a.m. the surf's increasing power
Still dashed with fury on the rocky shore—
Her solid timbers breaking with the strain
Sad presage gave, she ne'er would float again.
And while we watched with anxious beating hearts,
Expecting every sea the ship would part,
Three furious rollers, with resistless shocks,
Lifted her frame and dashed her on the rocks;
With awful crash her masts and timbers quivered,
And her whole hull was into fragments shivered.
Heart-rending, piercing, wailing cries were heard,
For all seemed over, and the worst was feared;
And then we thought no living soul was saved,
As through the watchfire's glare our eyes we strained,
And only listened if we could hear a cry,
Or could the motion still of life descry.
We saw the stern and poop still borne on the wave,
The means ordained by Providence to save;
And as we moved towards the place where each
Expected that the floating wreck would reach,
Piercing the flickering gloom, we ken
The narrow space crowded with living men.
With heartfelt cheers, but yet with dread, we hail
Their dangerous approach, lest still they fail;
And hundreds fly through the wild surge's roar
To help the helpless to our friendly shore.
These, the last remnants of that numerous band,
Were nearly all who landed on our strand.
A few more rolling waves, and nought was seen
Of that ill-fated ship, or where she had been;
The wild sea rolled, and furious surf broke on,
As if regardless of the mischief done.
And now the kindly feelings of the heart
Were called in exercise to do their part;
And women, foremost in the work of love,
To comfort and relieve their sufferings strove;
And happy they who have the means and will
The cup of love and charity to fill;
Far blessed, far more blessed 'tis to give,
Than from the purse of others to receive.
With many thus, the voyage of life was o'er.

If, perhaps, unprepared, they landed on the shore
From whence no voyager returns, to tell
Whether with him 'tis ill or well.
But this we know, each to his own place is gone,
The place he sought while here he landed on.
'Tis not our part their present lot to tell—
The Judge of all the earth will sure do well.
Let us who witnessed this sad scene take care
While on the voyage of life we thus prepare;
See that our cable and our anchor's good
To hold our cable when in Death's cold flood,
Then having weathered out the storm of strife,
On this we will enter on eternal life.

"H.M.'s 27th Inniskillings.

"Port Elizabeth, Algoa Bay, South Africa."

[I give the latter version of the account more on account of the value of the subject than that of the rhyme. The only soldiers of those saved serving in the regiment now are Feeney and McEnroe.]

I omitted to notice in its proper place that I find by the file of the *Fermanagh Reporter*, that the non-commissioned officers of the regiment entertained in the Royal Barracks, Dublin, "their respected sergeant-major, William M'Donald, previous to his leaving the regiment." After dinner Sergeant-Major Hurst, in a very suitable and appropriate address, in the name of the commanding officer and officers of the regiment, presented him with a richly-chased silver tea service, with an inscription. In the course of the evening Mr. M'Donald was also presented with a beautiful chased silver cup, from the sergeants, having on the lid the Castle of Inniskilling with a flag and an inscription. The sergeants of the 27th also presented the drum-major with a very handsome silver-mounted stick, on his leaving his regiment to join the English militia.

INDIA.

1854.—The first division, under the command of Major Tounzel, landed at Calcutta on the 20th September, 1854, and marched to Dum Dum. It embarked at Kossepore, and arrived at Allahabad. The head-quarters division, under the command of Lieutenant-Colonel Williamson, landed at Calcutta on the 12th October, 1854, and proceeded to Dum Dum the same night. This division embarked at Kossepore on board the river steamer, on the 16th October, and disembarked at Allahabad on the 19th November, 1854. The third division, under the command of Major Durnford, landed at Calcutta on the 26th October, 1854, and likewise proceeded to Dum Dum. It embarked on board the river steamer at Kossepore, and disembarked at Allahabad. The last division not having arrived, these three divisions commenced their march, under the command of Lieutenant-Colonel Williamson, from Allahabad on the 26th December, 1854, towards Sealkote, the regiment marching through Cawnpore on the 5th January, 1855, Delhi on the 2nd February, Umballah on the 15th, and was inspected on the 16th by General Sir William Gomm, the commander-in-chief, at Mogul-ki-Serai. Passed the Jullunder on the 27th February, 1855, crossed the Sutlej, passed through Umritsur on the 5th March, and arrived at Sealkote on the 14th March, 1855. The regiment was particularly healthy on the march, not having a single casualty.

The fourth division, under the command of Captain Stapylton, was wrecked at Port Elizabeth, Cape of Good Hope, on the 20th September, 1854, as before related. Captain Warren and Assistant-Surgeon Kidd were on board during the wreck, and displayed the greatest coolness and ability during this trying occasion, the other officers having landed the previous day for the purpose of providing provisions for the detachment, and were unable, owing to the heavy gale, to join the ship. Sixty-two men, eleven women, and twenty-six children were drowned, and the whole of the officers' baggage, arms, appointments, etc., and the vessel became a total wreck. The survivors were forwarded to Cape Town on the 13th October, on board her Majesty's steam sloop *Hydra*, where they arrived on the 16th, and embarked on board the ship *Maidstone* on the 9th November, 1854, for Calcutta, where they landed on the 9th

January, 1855, and were quartered in Fort William until the 13th, when they embarked on board the *Kallugunga Fah* for Allahabad, which place they reached on the 11th February. The detachment was forwarded in three divisions by Gort bullock train to Kurnaul, where they all arrived on the 27th February, and then marched to Sealkote, and joined head-quarters on the 31st March.

The regiment was inspected by Major-General G. E. Gowan on the 18th March, and again on the 7th November, 1855, by the same officer, who presented the following letter:—

"Head-quarters, Sealkote, 10th November, 1855.

"SIR,

"Major-General G. E. Gowan, C.B., commanding the division, having completed the half-yearly inspection of H.M.'s 27th Regiment under your command, has desired me to convey to you his entire satisfaction. The general appearance of the regiment under arms and in barracks, and its high state of discipline, do yourself, the officers, non-commissioned officers, and men of the regiment great credit, and it will afford the Major-General great satisfaction in reporting favourably to army head-quarters.

"I am, etc.,

C. PRIOR, Major,

"Deputy Assistant Adjutant-General, Lahore Division."

1855.—Lieutenant-Colonel H. D. Kyle joined and assumed command of the regiment on the 28th November, 1855. A draft consisting of 255 men, 39 women, and 28 children embarked off Gravesend in July, 1855, under the command of Lieutenant G. R. Gresson, and arrived at Kurrachee in November, 1855, and joined head-quarters by detachments as follows:—Lieutenant Gresson, Lieutenant Geddes,* 1 sergeant, 2 corporals, 184 privates, 34 women, and 21 children, on the 19th February, 1856. Lieutenant W. H. Davis, Ensigns Story and Stafford, 1 sergeant, 63 men, 8 women, and 11 children.

* Now junior major.

1856.—On the 1st April tunics were issued to the corps as new clothing (820), also new pattern dress caps (80). In the month of October, Nos. 5 and 7 Companies were furnished with new pattern accoutrements (double pouch) to replace those lost in the wreck of the ship *Charlotte* on the 21st September, 1854. On the 18th March, 1856, the corps was again inspected by Major-General G. E. Gowan, C.B., who expressed his entire satisfaction; *vide* regimental orders of 19th March, 1857 :—

"The Commanding Officer has great satisfaction in making known to the regiment that Major-General Gowan has expressed to him his entire satisfaction with the appearance and movements of the corps on parade, as well as with what has already fallen under his observation in barracks this day, and he will not fail to report favourably thereof to head-quarters."

On the 9th of October, 1856, the regiment lost Brevet-Lieutenant-Colonel G. A. Durnford, who died at Sumla when on leave, and who was deeply regretted by all ranks, having been thirty-one years and nine months in the corps. On the 14th November, 1856, the regiment was inspected at Sealkote by Brigadier F. Brind, C.B., who forwarded the accompanying letter to Lieutenant-Colonel H. D. Kyle, commanding :—

"Sealkote, 4th November, 1856.

"Sir,

"On the occasion of her Majesty's 27th Regiment being paraded for inspection this morning, I find all so good that there is little room to particularize, and I can but regret that yourself and regiment had not the advantage of the discrimination of the General Officer commanding the division; for I make no question that, in addition to the unqualified satisfaction I am enabled to express, much deserving of remark would have attracted his favourable notice.

"2. It appears to me that the field movements were executed with great accuracy and precision, and that the regiment was particularly steady and correct in its marchings, its wheeling either by company or column, and its other changes of position. Distance, that great test of attention and instruction of the commissioned ranks, was correctly preserved throughout, and in all changes and formations the several portions of the regiment, as they successively formed on the alignment, were close and compact, well placed and prepared to act, acquirements of the highest importance when before the enemy.

"3. Adverting to the recent severe sickness and to the drill season having but recently commenced, I did not expect field movements, but in those I viewed there was proof of the attention that has been paid to the instructions received from army head-quarters for keeping up regimental efficiency during the year.

"4. The appearance of the regiment and its performance satisfy me that it is at this time in a high state of efficiency for any service in which it might be employed, and the evident desire on the part of all to contribute to this satisfactory result must be highly gratifying to you, as it was pleasing to the inspecting officer.

"5. I request you will make known my sentiments in such manner as you may judge best.

"I have, etc.

(Signed) "F. Brind, C.B.,

"Brigadier commanding Sealkote Brigade.

"To Lieutenant-Colonel H. D. Kyle,
commanding 27th Regiment."

The regiment marched out of Sealkote on the 15th November, 1856, *en route* for Nowshera, which station was reached on the 9th December, 1856.

On the 25th December, 1856, the regiment again marched towards Peshawur, to form part of the force ordered to be present at the interview of the chief commissioner and the Ameer of Cabul, and took part in the grand review before Dost Mahomed on the 31st December, near the Khyber Pass.

Major-General T. Reid, C.B., reviewed the regiment on the 6th January, 1857, and declared himself highly pleased at the steady and soldier-like appearance of the regiment in the marching and in the field movements. The regiment returned to cantonment at Nowshera on the 11th February, 1857.

1857.—On the 22nd March a detachment from the regimental depot arrived at head-quarters—strength one captain, one lieutenant, one ensign, and 108 men.

I find in a marginal note that those officers were Captain Freer, Lieutenant Simeon, and Ensign Attinwood.

The annual clothing (shell jackets) was issued to the corps on the 1st April, 1857.

On the 7th April, 1857, the regiment was inspected by Major-General T. Reid, C.B., commanding the Peshawur division, *vide* regimental orders, 9th April, 1857 :—" Lieut.-Colonel Williamson has much pleasure in publishing that he has been commanded by Major-General Reid, C.B., to communicate to the Inniskilliners 'the great satisfaction it afforded him in witnessing, at his inspection of yesterday, the very clean and soldier-like appearance of the regiment, as well as the steadiness and precision with which they performed the various battalion and light infantry movements.' "

INDIAN MUTINY.

On the 14th May, 1857, intelligence was received of the mutiny of the native army at Meerut, and instructions were issued to hold the regiment in readiness to march, fully equipped for service, to Sheelum to join the moveable column. The corps marched the following day under command of Lieut.-Colonel Kyle, the strength being two field officers, two captains, seven lieutenants, four staff officers, fifty sergeants, forty-one corporals, eighteen drummers, and 817 privates. On arrival at Hussun Abdul on the night of May 18th, orders were received by express from division head-quarters, directing the head-quarters and right wing to return and occupy the fortress of Attock, on the left bank of the Indus, the left wing, under command of Lieutenant-Colonel Williamson, proceeding on to Rawul Pindee. The order was duly carried into effect, each wing reaching its destination. On the 22nd May three companies of the left wing, under command of Captain Warren, made forced marches on elephants and gun carriages from Rawul Pindee to Attock, arriving on the morning of the 24th May, and relieved the right wing and head-quarters of the corps, who (as the 55th Native Infantry had broken into open mutiny at Murdoon and Nowshera, where the soldiers' families and heavy baggage, etc., had been left in charge of Lieutenant Davis) proceeded the same night, under command of Lieutenant-Colonel Kyle, by a forced march to Nowshera, crossing the Indus in boats. Arrangements were immediately made on arrival for despatching the women and children to Rawul Pindee.

On the night of the 27th May the head-quarters and right wing having made a forced march into Peshawur, was brought on the strength of the garrison and quartered in the lines of the 70th Foot, detachments being furnished to the Fort Machison's Post, besides various picquets. On the 6th June, 1857, the regiment furnished three sergeants, one corporal, fifty privates, as volunteers to the Peshawur Light Horse, a local corps raised by General Cotton from volunteers from the queen's troops in the division. On the 8th June the remaining two companies of the left wing marched from Rawul Pindee to Attock, leaving the sick women and children (who had joined from Nowshera) at Rawul Pindee, in charge of Lieutenant Geddes.

On the 15th June the left wing, under command of Lieutenant-Colonel Williamson, marched from Attock to Nowshera, arriving the same night. Two companies (Nos. 7 and 8) were despatched on the 29th June, under command of Captain Langley, to Peshawur, to reinforce the head-quarters. On the 8th July an order was received to move three companies (5, 6, and light) from Nowshera to Attock. Lieutenant-Colonel Williamson marched the same evening, and arrived under the walls of the fortress at daybreak on the 19th inst. The heavy baggage and the regimental stores, etc., were left at Nowshera until the end of August, in charge of a guard (thirty-two men).

On the 28th July a detachment, consisting of two sergeants, two corporals, one drummer, and forty-six privates, was ordered into the Ensuzaie country to form part of the column sent under Major Vaughan (5th P.I.) against the village of Naringee. Assistant-Surgeon Kidd accompanied the party, and Captain Barnes commanded the European division, formed of detachments from H.M.'s 70th and 87th Regiments. The village was stormed without casualties amongst the Europeans, and the force rejoined at Peshawur on the 11th August. On the 3rd August the regiment furnished one corporal and twenty-eight privates as volunteers to the 4th Troop, 2nd Brigade Bengal Horse Artillery. Lieutenant H. B. Patton was also attached to the troop.

On the 28th August the regiment was present and took part in the entire destruction of the 51st Native Infantry, which mutinied *en masse* at Peshawur. The heat was excessive, but all ranks behaved with great steadiness, and officers and men were thanked by the general commanding the division. On the 9th October, 1857, Lieutenant-Colonel Williamson's detachment (5, 6, and light) marched from Attock to join the head-quarters at Peshawur, and joined on the 11th inst.

Lieutenant-Colonel H. D'Arcy Kyle died at Peshawur, on the 11th October, 1857, from fever with dysentery. The decease of this most gallant and inestimable officer, who so fully earned the confidence, love, and esteem of *all* ranks, was severely felt by the regiment. Lieutenant-Colonel Williamson assumed command of the regiment.

The severe duty, continued exposure, arduous night-work, was so cheerfully and efficiently performed by all ranks, that it called for the highest eulogiums from the major-general commanding. The health of the regiment was, however, much impaired. The corps lost by death during the year 147 men. The regiment returned to Nowshera on the 1st December, 1857.

1858.—On the 13th February, 1858, 160 men, with their families, 2 subalterns, 1 assistant-surgeon, under command of Captain Freer, were sent to Rawul Pindee, for change of climate. On the 16th February, 1858, a detachment of recruits joined from England—Captain Croker, Lieutenant (now Captain) Caine, Lieutenant Desborough, Ensigns Dixon and Clay, and 35 privates. On the 15th March, 1858, the regiment marched from Nowshera for Umballa. Captain Freer's detachment joined at Rawul Pindee.

On the 16th 700 Enfield rifles had been issued to the corps just before leaving Nowshera. On the 7th April Captain Mitford and Lieutenant Surman, with 54 recruits, joined the regiment at Meanmeer on the line of march. The regiment marched into Umballa on 25th April. On the 16th May 16 recruits joined from England.

The regiment furnished detachments at the hill stations of Kussowlie and Dugshin while quartered at Umballa. The annual clothing (single-breasted tunics) for 1858 received at head-quarters in the month of August, and was fully issued by 1st September.

Colonel Williamson, from ill health, was obliged to return to England, and handed over the command of the corps to Captain B. Thomas (on the 7th August, 1858), being the senior officer present. This officer retained command until 15th April, 1859. The regiment was inspected by Major-General Sir Robert Garrett, K.C.B., on the 15th October, 1858.

MUSKETRY.

The annual course of musketry instruction, under Lieutenant G. S. White (now Captain White), commenced on the 16th October, and concluded the 1st April, 1858. Fifty-one men qualified themselves as marksmen; No. 2727, Private William Day, No. 2 Company, as the best shot of the battalion. The figure of merit was 29-10. In November 3 sergeants, 1 corporal, 1 drummer, and 45 privates were invalided to England. New pattern belts, with double pouches, were issued to the regiment in February, 1859. Lieutenant-Colonel Stapylton joined from leave on the 15th April, and took over command of the corps. Wicker-work helmets were issued to the regiment in the month of April, 1858.

The regiment was inspected by Lord Clyde, the commander-in-chief, on the 25th April, 1858. His Excellency expressed himself much pleased with the steadiness of the men under arms and the general discipline of the corps. 219 recruits joined from the depôt since May, 1858; also Ensign Cobbe, who arrived December 26th, 1858.

1859.—The regiment was again inspected by Lord Clyde, the commander-in-chief, on the 9th October, 1859. On the 14th November 2 sergeants, 3 corporals, and 37 privates were invalided to England, and 38 privates and 1 corporal obtained free discharges.

On the 17th November Ensigns Hamilton and Tottenham joined from England. Lieutenant Campbell died *en route* to join from the regimental depôt. The annual clothing was issued on the 1st September, 1859, serge tunics being served out instead of jackets. On the 17th November the regiment was inspected by Major-General Sir R. Garrett. Lieutenant-Colonel R. S. Baumgartner, C.B., was appointed to the regiment on the 22nd July, 1859, and joined and assumed command at Umballa on the 20th April, 1860.

At the course of musketry Private Andrew M'Donnell qualified himself as the best shot of the battalion. Major-General Sir R. Garrett, C.B., inspected the regiment on 2nd November. He "approved of their general good conduct," and said their appearance was that "of a very clean and fine body of men, steady under arms, and very close and compact in their movements, particularly in their advance in line."

1860–61.—The regiment marched from Umballa on the 10th November for Morar Gwalior, and arrived on the 16th December, 1860. With the draft of 135 men from the regimental depôt, which arrived on the 21st January, 1861, came Ensign (now Captain) Herring. The regiment was inspected by Brigadier J. Welchman, C.B., on the 20th March, 1861, at Gwalior.

The regiment suffered severely from cholera during the months of July, August, and September, 1861, whilst quartered at Gwalior. The deaths resulting from this epidemic amounted to 193, besides two officers. The establishment of the 27th Inniskillings was reduced by War-Office letter to a total of 1079.

The regiment marched from Gwalior on 26th December, 1861, for Gondah in Oude. On arrival at Lucknow on 16th January, 1862, the battalion was halted, was inspected the 20th January by Brigadier-General Renny, C.B., commanding the Oude division, continued its march on the 26th January, and arrived at Gondah on the 1st February. The regiment was inspected by Major-General MacDuff, C.B., commanding Oude division, on the 15th April, 1862, and again on the 12th January, 1863. The regiment was inspected at Gondah on 14th February, 1863, by his Excellency General Sir Hugh Rose, G.C.B., commander-in-chief in India, who expressed himself much pleased with the appearance of the regiment.

The regiment marched from Gondah for Dinapore on the 2nd November, 1863, and arrived at the latter place on the 11th December, 1863. Major-General Sir S. Corbett, K.C.B., inspected the regiment on the 22nd December, 1863, and the 16th March, 1864.

Ensign Stainforth (now lieutenant) joined from England on the 4th June, 1864.

On the 28th April, 1864, at Edinburgh, died Lieutenant-General John Geddes, K.H., colonel of the 27th Inniskillings. This officer served 20 years in the regiment, and was present at the battles of the Nivelle, the Nive, Orthes, Toulouse, at which last he received a severe wound which broke his left thigh-bone.

[General Geddes was uncle to the present junior major of the 27th.]

Major-General J. M. Crawfurd was appointed colonel of the regiment. Ensign (now Captain) White joined the regimental head-quarters in the month of September, 1864. Brigadier-General Colin Troup inspected and complimented the battalion on November 22nd, 1864; and on January 16th, 1865, Lieutenant-General Sir Hugh Rose published the following remarks upon his inspection on that day:—"The Lieutenant-General has much gratification in communicating his unqualified approbation of everything connected with the interior economy of the Inniskillings. . . . ."

In the beginning of March, 1865, the 20th Regiment relieved the Inniskillings at Dinapore, and the regiment was ordered to march to Hazareebagh, detaching two companies to Berhampore to relieve two companies of the 55th Regiment, in command of Major Freer. Lieutenant Geddes was one of the officers of this party.

The regiment arrived at its destination on Saturday, 25th March, 1865, and on the 7th April, 1865, the regiment was inspected by Major-General Sir S. Corbett, K.C.B. The annual course of musketry for 1865-66 was concluded on the 12th May, 1865, under Ensign White, M.I. On the 18th August Colonel Baumgartner left for England on sick leave, and resumed command on the 22nd March, 1866. The regiment was inspected by General Welchman on the 5th April, 1866, and again on the 9th November.

On the 27th December, 1867, the regiment marched from Hazareebagh, *en route* for Dum Dum, and arrived on 13th January, 1867. On the 1st October, 1867, the head-quarters and detachments of Barrackpoor and Berhampore moved into Fort William, Calcutta.

On the 13th November the regiment embarked at Calcutta on board H.M.S. *Euphrates* for Suez. It arrived on the 5th December, but remained on board the troop-ship until the 20th December, on which date the disembarkation took place. The regiment proceeded thence by rail to Alexandria, arriving at the latter place on the 27th of December, at about four p.m. The sea being considered too rough to proceed with the re-embarkation, the officers and men remained for the night in the railway carriages, and embarked on board H.M.S. *Crocodile* at seven a.m. next morning.

HOME AGAIN—PORTSMOUTH.

1864.—On the 13th January, the 27th Inniskillings, under command of Major Baumgartner, C.B., landed at Portsmouth, and proceeded thence by rail to Dover, where the depôt of the regiment was stationed at the time. On arriving at Dover the regiment was ordered to occupy the Citadel Barracks on the western heights. On the 15th January the regiment was inspected by Major-General Ellice, C.B. On the 17th January the service and depôt companies were amalgamated and formed into ten companies. On 2nd June the regiment was inspected by Colonel Kirkland, 2nd Battalion 5th Fusiliers, commanding the garrison. On the 15th June the regiment, under command of Major R. Freer, was inspected by Major-General Ellice, C.B., commanding South-eastern Division, who expressed himself entirely satisfied with the appearance of the corps.

On the 22nd June the left wing of the regiment, under command of Major Murphy, proceeded to Shorncliffe Camp, and were quartered in B lines. On 23rd June the regiment furnished a part of 126 rank and file, for employment on the works at Castle Hill Fort, Dover. On 8th October the regiment, under command of Major Freer, was inspected by Major-General Russell, C.B., commanding South-eastern Division. The general expressed himself well satisfied with the appearance of the corps on parade, and also with the interior economy. On 16th October the detached companies of the regiment rejoined the head-quarters from Shorncliffe Camp and Castle Hill Fort. On 7th November the regiment furnished two companies to Isle of Grain Fort, under command of Brevet-Major O'Donnell; one company to Tilbury Fort, under command of Captain Cowell; and one Upnor Castle, under command of Lieutenant Brownrigg.

CHATHAM.

The remaining companies and head-quarters of the regiment, under command of Colonel Baumgartner, C.B., moved to Chatham from Dover, and were quartered in St. Mary's barracks on the 10th November. On the 19th November the head-quarters companies of the regiment were inspected by Major-General Freeman Murray on the Chatham lines. The general was satisfied with the appearance of the regiment.

1869.—On the 14th May, 1869, the head-quarters companies were inspected by Major-General Freeman Murray, who expressed himself pleased with the appearance of the men on parade, and the extremely clean state of the barrack-rooms. On the 21st May the detached companies rejoined head-quarters at Chatham. On the 13th July 205 Mutiny medals were issued to the officers and men.

NEW COLOURS.

I need scarcely observe that no person was more fitting to make the presentation than the Countess of Enniskillen, nor could it have been made in more graceful terms. The Earl of Enniskillen, mindful of the traditional memories of his house, has ever manifested the greatest interest in the county regiment, and it was meet that the wife of the representative of a family who assisted at the regiment's institution (and whose uncle was a colonel of it) and lord of the soil of the town from which it takes its name, should have the honour of presenting the new colours.

On the 19th July, 1869, the regiment paraded on the great lines at Chatham (about 500 strong), under command of Lieutenant-Colonel R. Freer, for the presentation of new colours, which was made by the Countess of Enniskillen. On this occasion the men appeared in the new pattern clothing of bright scarlet colour. Due honours having been paid to the old colours, Major-General Freeman Murray, the Earl and Countess of Enniskillen, staff, and spectators advanced to the saluting flag, the regiment forming three sides of a square. The new colours were placed leaning against a pile of drums in the centre. The Rev. T. J. Coney, M.A., chaplain to the forces, performed the religious part of the ceremony.

The Countess of Enniskillen now presented the new colours to Ensign H. Wodehouse and C. W. Hare, these officers kneeling. Her ladyship said she felt deeply the honour conferred upon her in being permitted to present these colours to so distinguished a regiment, closely related, as it was, with a part of the country with which she and hers were connected—a regiment which was raised 180 years ago in her husband's native town. She referred to the services of the regiment in many fields during the present century, and in the crowning victory of Waterloo they carried their colours triumphantly to the end. Their services had not been less valuable since, though the actions they had taken part in for the last fifty years were less known to fame. As the older soldiers of the corps had been succeeded by younger ones, so the old colours were to be succeeded by new ones. Her ladyship delivered them to their charge, praying that the Great Ruler would give them strength to bear them boldly, bravely, and steadily, and maintain the high character of the corps in times of peace.

Lieutenant-Colonel Freer thanked her ladyship for the kindness and honour she had done them in presenting these colours. The regiment then marched past, bearing their new colours, and subsequently returned to their quarters in St. Mary's barracks.

On the 29th September, the regiment was inspected by Major-General Freeman Murray (on the great lines), who was much pleased with the appearance and steadiness of the men on parade. He also found the barrack-rooms in excellent order.

1870.—On the 19th February I Company was sent on detachment to Upnor Castle.

COLCHESTER.

On 24th March the regiment marched from Chatham to Gravesend, and thence by rail to Colchester, relieving the late 4th and 8th Battalions. On passing through Brampton barracks the corps received a marked compliment from the Royal Engineers, who were drawn up in line under Colonel Gallwey, the commandant, each company being saluted with cheers. I Company, under Captain Swanton, joined the battalion at Rochester bridge. On 28th March, K Company, under Captain Davis, marched to Landguard Fort.

On 12th April the regiment paraded on the Abbey Field,

under Lieutenant-Colonel Freer, for the inspection of Major-General Freeman Murray, who expressed himself pleased with the state of the corps, both in camp and on parade. The depôt 81st L.I. Regiment joined the Inniskillings on the 31st March. The establishment of the 27th Regiment was ordered to be increased by 300 privates on 24th August. Sergeants were sent to recruit at Dublin, Belfast, and Armagh. Four hundred recruits joined between this date and February, 1871 (over 300 from the eastern counties). On 6th October the regiment, under Lieutenant-Colonel Freer, was inspected by Major-General Freeman Murray, who was highly pleased with the state of the regiment and the appearance of the men on parade. A Company, under Captain Simeon, rejoined head-quarters on the 7th October from Landguard Fort.

1871.—C Company, under Captain Cotton, proceeded to Landguard Fort.

THE 27TH "FIRST IN ORDER OF MERIT."

The excellency of the 27th Regiment as the best shooting regiment in the service for two years is *unequalled* in the annals of the British army. A regiment holding first place for two years and second place in the third year is a unique fact. This gratifying result was chiefly owing to the exertions and excellent training of Captain Esmonde-White, then musketry instructor.

On the 11th April the following letter was received from the adjutant-general:—

"SIR,

"With reference to the return (annexed to General Order No. 33 of the 1st April, 1871) showing the corps in the United Kingdom, etc., which had concluded a course of musketry instruction for the year 1870-71, I have the honour, by desire of the Field Marshal Commanding-in-Chief, to say that his Royal Highness has not failed to perceive that the 27TH FOOT IS THE 'FIRST IN ORDER OF MERIT,' *which position the regiment occupied in the previous year's course* of instruction;* and I am accordingly instructed to request that you will be pleased to cause the Commanding Officer and the Musketry Instructor of that corps to be informed that its efficiency in this respect is most satisfactory, and reflects the greatest credit on them, and indeed on all concerned, showing, as it does, the interest which must have been taken in carrying out this most important branch of a soldier's training. His Royal Highness feels sure that exertions will continue to be made to maintain the high position the 27th Foot now occupies in the musketry records of the army.

(Signed) "RICHARD AIREY, A.G.

"To General Officer commanding at Colchester."

On the 16th April C Company, under Captain Cotton, rejoined head-quarters. On the 2nd May the regiment was inspected by Major-General Freeman Murray on the Abbey Farm Field. He expressed himself well pleased with all he had seen of the regiment in quarters and on parade.

ALDERSHOT.

On 31st May the left wing, under Major Cowell, proceeded by special train to Aldershot. The following day the head-quarters, remainder of the regiment, and depôt 81st Regiment, under Lieutenant-Colonel Freer, arrived at Aldershot, and encamped near the North Camp Church. The regiment was attached to the 3rd Infantry Brigade, commanded by Major-General G. V. Maxwell, C.B. During the autumn manœuvres the regiment acted with the 2nd Division, under Major-General Carney, remaining under canvas until the 29th September, when it moved into the huts vacated by the 2nd Battalion 9th Regiment in the North Camp. On 6th October the regiment was inspected by Major-General Maxwell, C.B.

1872.—On 17th April the regiment was inspected by Major-General Maxwell, C.B. By army circular of 1st May the establishment of the regiment was reduced to 48 sergeants, 40 corporals, 18 drummers, and 780 privates.

* The small capitals and italics are my own.

GOSPORT.

On the 21st August a detachment under Captain Geddes, together with the depôt 81st Regiment, left Aldershot for Gosport, and arrived at the new barracks the same day. On 26th August the regiment, formed into eight companies, joined the division under Lord Mark Kerr, and was brigaded during the autumn manœuvres with the 2nd Battalion 4th "King's Own," and South Gloucestershire Militia, under command of Colonel Pakenham, 30th Regiment. On 18th September the regiment arrived at Gosport new barracks.

1873.—On the 10th February G Company, under Captain Phillips, proceeded on detachment to Upnor Magazine. On 5th March the depôt 81st Regiment, which was attached to the regiment since the 31st March, 1870, joined the 1st Battalion 4th King's Own Regiment.

CURRAGH, IRELAND.

On 7th May the regiment embarked in H.M.S. *Himalaya* for Ireland, and arrived at the Curragh Camp on 20th May. By army circular of 1st July, 1873, the establishment of the regiment was reduced to 42 sergeants, 40 corporals, 18 drummers, and 480 privates.

MEETING OF THE TWO INNISKILLING REGIMENTS.

Whether it falls within the province of the Records or not I cannot say, but I know I find no mention made of the cordial greetings between the 27th Foot and the 6th Horse Inniskilling Regiments at the Curragh Camp. More trivial events have been noted. These two distinguished regiments had not met since the field of Waterloo, where both played important parts. Above all others, the colonel and sergeant-major of a regiment should be "good men." The 27th has both: Sergeant-Major Windrum possesses the *esprit de corps* in such measure as to inspire the non-commissioned and subordinate ranks with it, and I need hardly add the result therefrom will be corresponding good conduct. When the notification of the entry of the 6th was heard, Mr. Windrum got permission from Colonel Freer for the use of the 27th's band. It is a point of military etiquette that cavalry bands shall play cavalry regiments into or from stations, while foot bands do the same for foot regiments. There was a slight difficulty here which Mr. Windrum got rid of. He went to the colonel of the 4th Dragoon Guards, and requested that the 6th Dragoons should be brought by a roundabout to their quarters so that they should pass through the lines of the 27th, and that the band of the 4th Dragoons should give way to the band of the 27th when the 6th Inniskillings entered the lines of the 27th. The request was granted, and Colonel Thesiger, of the 6th, acquiesced in the arrangement. Thus was the programme carried out. The famed Inniskilling horse was led up to the lines of the Inniskilling foot, as the 27th's band played "Inniskilliners round the Globe." The 27th had turned out and formed two lines, and between the long avenue of their Inniskilling brethren were the famed soldiers of the 6th escorted to their quarters amid the heartiest cheering. The 27th on that evening, with their accustomed hospitality, entertained the 6th Dragoons, and after fifty-eight years of separation the two Inniskilling regiments renewed friendly ties, linked as they were in one common chain of brotherhood. The dragoons returned the hospitality on a subsequent occasion. The following pleasing letter afterwards appeared on the regimental orderly books of the 27th Regiment, dated 7th June, 1873:—

"The Lieutenant-Colonel commanding has been requested by Hon. Colonel Thesiger, commanding 'the Inniskilling Dragoons,' to express to the regiment the warm appreciation of the dragoons of the hearty reception with which they were met on arrival at the Curragh this day by 'the Inniskilling Infantry,' and which, Colonel Thesiger is kind enough to add, will not be forgotten."

1874.—On 23rd April a detachment under Major Cowell, consisting of B and D Companies, left the Curragh camp for Londonderry, and arrived the same day.

During their stay in the Curragh the sergeants presented Quartermaster-Sergeant C. Brennan with a sword and belt, as a mark of their esteem, on his appointment as quarter-master of the regiment. The presentation was made by Sergeant-Major Windrum, the vice-chair being filled by Quartermaster Holles.

ENNISKILLEN.

A detachment under Captain Herring, consisting of F and E Companies, left the Curragh for Enniskillen, and arrived the same day. By army circular of 1st May the establishment was increased by forty rank and file. A Company, under command of Captain White, left the Curragh for Newbridge, there to be employed on the engineer's works, on 11th May. On 8th July the head-quarters of the regiment, under command of Colonel Freer, with depôt of 108th Regiment attached (27th consisting of seven officers, and 180 rank and file and sergeants; 108th, four officers, 93 rank and file, and four sergeants), left the Curragh for Enniskillen.

After an absence of twenty-one years the head-quarters of the regiment, under command of Colonel Freer, arrived at Enniskillen from the Curragh Camp on the 8th July, 1874. A large number of townspeople assembled to witness tho arrival of the regiment, and to demonstrate their welcome. Triumphal arches and flags were displayed in the principal streets of the town, the church bells rang, and the guns from the Fort hill fired salutes as the regiment marched through the town and barracks.

A detachment under command of Captain Phillips, consisting of C and G Companies, remained for duty at the Curragh Camp. On 30th July C and G Companies, under command of Captain Phillips, left the Curragh Camp for Royal Barracks, Dublin, there to be quartered. On 8th July F Company, under command of Captain Herring, left Enniskillen for Londonderry, arriving at the latter place the same date.

THE TWO INNISKILLING REGIMENTS IN ENNISKILLEN.

Wednesday, the 15th September, 1875, was an historical day in the annals of the two Enniskillen regiments. For the first time since the seventeenth century, a party of the 6th Inniskilling Dragoons met the 27th Inniskilling Foot in the island town which gave birth to those two regiments and the 5th Royal Irish Regiment of Dragoons. Well-nigh 200 years have passed since two troops of Sir Albert Cunningham's dragoons and Tiffin's foot regiment returned to Enniskillen from the capture of Belturbet and Cavan; and from that time till the day afore-mentioned, in this year of our Lord 1875, they have not marched together up Enniskillen streets. Side by side they fought and conquered at the Boyne; together they fought and were victorious at Limerick. Under their sovereign in Flanders they supported his cause. Again we find them together in Germany, again in Flanders; and in 1815 together in the field of Waterloo, where both regiments were distinguished by their bravery. Fifty-nine years elapse, and the 6th Inniskilling Dragoons enter the Curragh Camp through the lines of the 27th Inniskilling Foot, amid deafening cheers; and both regiments again renew the ties of affection which have bound them together since the 1st January, 1690, the day on which King William III. "thought fit to forme . . . . two regiments of dragoons and three regiments of foot out of our Inniskilling forces."

Throughout those wars in which the 6th Dragoons took part, the men of Enniskillen and Fermanagh maintained the good name merited by their forefathers. It still is told by many families in the county that their ancestors fought in the Inniskilling Dragoons on a foreign field. The Prices of Belnaleck supplied the regiment with men of their name for a long and unbroken period. From the days of the Boyne they were sergeants-major and non-commissioned officers in the 6th Inniskillings. At the time of Waterloo there were twelve or thirteen of the name in its ranks; and the last man of the name found on the roll of the regiment, a brother* of the present Mr. Alexander Price, was one of those who took part in the terrific charge at Balaclava.

* The following is a copy of a letter written by Joshua Price, one of the Inniskilling Dragoons, while serving in the Crimea, to his brother, Alexander Price, Master of the Enniskillen Workhouse.

"Camp before Sebastopol, 28th December, 1854.

"DEAR BROTHER,—After a very long silence I send you a few lines, merely to give you a slight idea of our encampment, which is a very weary one to us; but you will get a better description of it in the newspapers than I can give. Ever since we landed in this country, we turn out in marching order at five o'clock a.m. and wait until the field officer goes his rounds and sees that all is right, and then turn in. On the morning of the 25th of October we had just turned in and unbridled, when the alarm sounded; in about three minutes every man was mounted and off at full gallop to meet his enemy. Our troop of Royal Horse mounted the hill in front of us. Then the play began. The shells from the enemy came over the hill, lit and burst amongst us, but luckily we had none hurt. We had three batteries on the hill betwixt the enemy and us, manned by the Bono Johnnies (that is, the Turks). The enemy threw a shell into the first battery; the Turks fled directly, leaving one of our gunners, who stopped till he spiked the gun, and then cut for his life. I was speaking to him after he came down the hill.

"The Russians very soon mounted the hill and took possession of the battery, then very soon had the second and third. We were obliged to retire out of the range of their guns. Then, in a few minutes, about three thousand of their cavalry charged on the 93rd. They stood their ground like men, and waited till they came within about four hundred yards; then they opened fire upon them, supported by the batteries in rear of them, which soon made them retire with great loss of life. We had not time to enjoy the spree until they poured down on us. We were in a very bad position at the time; they formed up in front of us, and outflanked us: we had not more than ten yards to go in at them. Our second squadron made the first attack—we were very soon alongside of them; it was then you would have heard the clash of swords. Our swords sprung off their great grey coats, but we put the points to work, which very soon entered them; and in a very short time the field was covered with grey coats. It would have pleased you to have heard the Inniskilling and Scots Greys call on each other as they were going into action.

"It was hard work for a short time, but we soon put them to flight. I received a slight wound by having my right ear cut through with the point of a sword, but it did not signify—I had revenge for it. We are still under canvas, so I leave you to judge the comfort we must have. If you were to see us now, you wouldn't think we ever belonged to the Inniskillings; our scarlet coats are black ones, and our overalls, you would not know what colour they are for mud and dirt. It was two months on the 17th of this month since we opened the siege on Sebastopol, and never gave over day or night, and no sign of it being taken yet. We are expecting another engage-

The head-quarters of the dragoons lying at Dundalk, and the head-quarters of the foot at Enniskillen, Lieutenant Edward Barton, the only Fermanagh gentleman in the 27th, invited the Hon. Colonel Thesiger, officers, and men to his residence at Clonelly. Mr. Motte, Mr. Waldron, Sergeant-Major Adams, and a party of non-commissioned officers and men represented the dragoons on the occasion. They were received at the railway station by Colonel Freer, the band, and a guard of honour; and by the townspeople with warm acclamations and ringing of bells. A most enjoyable day was spent at Clonelly; and having enjoyed the hospitality of the 27th Regiment for another day, the party returned, regretting their quarters were not in the old town.

PRESENT STATIONS.

The regiment is at present distributed thus:—Head-quarters, Enniskillen, C and G Companies (with depôt of 108th Regiment attached), under command of Colonel R.

ment every day. Brother Jason's two sons are here, Robert and Alexander; they have been with me several times. Alexander got a wound on the 5th of November by a ball; the last time I heard from him he was getting on as well as could be expected. Robert is all right. I hope this will find you and brother Robert in good health, and think yourself a happy man that has a house to cover your head, and knows not the hardships of this country. Give my kind respects to Mrs. Carson and daughter, and to Mrs. Britton, and tell them I wish I was alongside of them, but such a sight I never expect to see. You must excuse this writing, for I have written it on the bottom of my mess tin. I wish you would be so kind as to send me a few stamps in your letter, for there is no possibility of getting any here. I must conclude, for my paper is out, in hopes that we shall meet again.

"I remain your affectionate brother,

"JOSHUA PRICE.

"(Direct to Joshua Price, 6th Dragoons, serving in the British Army, Crimea.)

"To Alexander Price, Master of the Workhouse, Enniskillen.

"P.S.—I have kept this for several days, and could not get a stamp for it; let me know what it will cost you.

"I hope you will not neglect to answer this. God only knows, I may be in eternity before that time, yet I know the Lord is able to bring me through all I have to contend with."

Freer. At Derry—D, E, F, H, and K Companies, under command of Major H. Cowell. At Boyle—A and I Companies, under command of Captain R. W. E. White. At Sligo—B Company, under command of Captain F. Coffey.

PRESENTATION TO QUARTERMASTER-SERGEANT WINDRUM.

1875.—On Friday evening, the 2nd April, a pleasing presentation to Quartermaster-Sergeant Windrum, late hospital sergeant of the 27th Inniskillings, took place in Enniskillen Town Hall, where the sergeant-major and sergeants of the 28th Inniskillings had a number of guests to witness the ceremony. Colonel Freer presided on the occasion.

And thus will we leave them, uttering the wish that the same honourable and glorious career may ever remain a remarkable characteristic of her Majesty's 27th Inniskillings.

Hail, Enniskillen! let thy fame
Emblaze thy country's story,
And every tongue confess thy name,
Synonymous with glory.

APPENDIX.

BATTLE OF NEWTOWNBUTLER.*

After the junction of Colonel Berry and Colonel Wolseley at the moat above Lisnaskea, Colonel Wolseley sent on the forlorn hope about half a mile before his army against General Macarthy. Colonel Tiffan (afterwards head colonel) led the first battalion of foot, consisting of about five or six companies, supported by a few troops of horse. Colonel Lloyd commanded the second battalion of foot, consisting of nearly the same number, seconded in a similar manner by cavalry. The main body of foot was led on by Colonel Wolseley himself, followed by the rest of the horse, under the command of Lieutenant-Colonel Berry and Major Stone.

In this order they marched from Lisnaskea to Donagh, through which they passed, and within half a mile of it got in view of the enemy's forlorn. About the same distance from Newtownbutler they discovered the Irish army posted very advantageously on a steep hill commanding a long and narrow causeway through a bog, by which way only it could be approached from that side. The Enniskillen army, however, advanced against them with steadiness and vigour. Colonel Tiffan, with his battalion of foot, entered the bog on the right hand of the causeway, while Colonel Lloyd, with the body under his command, pushed on in the same direction on the other side. Colonel Wynn's dragoons, divided into two equal parts, supported Tiffan and Lloyd on foot. Lieutenant-Colonel Berry advanced at the same time on the causeway with his horse, Colonel Wolseley bringing up the main body in the rear, to send reinforcements to those who went before him, as occasion should require. In the mean time, the

* Selection from Rev. John Graham's History.

enemy very injudiciously exhibited a proof that they thought their position untenable by setting the town of Newtownbutler and the houses in its neighbourhood on fire. After a weak opposition the Enniskilleners gained the pass, and pursued them through Newtownbutler, and near a mile beyond it. The retreating army fell back in good order, and again took a position similar to the last one they had occupied, securing the narrow causeway leading to it by a piece of cannon. The pursuing army, making the same disposition as before, found the passage of their horse impeded by the fire of the cannon till the foot, advancing by the bog on each side, killed the cannoneers, and rushed on towards the enemy in the hill, upon which the Irish horse took fright, and flew towards Wattlebridge, deserting their foot.

The foremost in this disgraceful flight was Lord Clare's regiment of horse, called the Yellow Dragoons from the colour of their facings. The tale of their dishonour is yet told in the barony of Moyarta, near the mouth of the Shannon, where they had been raised. It is told in the way of dialogue, in which a person supposed to have witnessed the scene says, "Stop, stop, Yellow Dragoon!" to which one of them replies, "Not till I get to the bridge of Clare!" another, "No, no, not till we come to the ford of Moyarta." Captain Martin Armstrong, with a troop of cavalry, did great execution on these fugitives. The Irish infantry, now abandoned by their horse, and closely pursued by the Enniskilleners, fled into a large bog towards Lough Erne on the right hand, throwing away their arms into the turf pits as they went. An open country lay upon their right, through which they might easily have escaped, but with their usual want of presence of mind it did not occur to them to prefer it. They were followed by the Protestant foot through the bog into a wood near the lough, where, no quarter being given to any but officers, 500 of them took to the water, and of these *only one man* escaped drowning: he got away safely by good swimming, though many shots were fired after him. During the whole of this night the pursuers were beating about the bushes for the Irish, and their officers were unable to recall them until next morning, by which time scarcely a man who had fled from them into the bog escaped death.

There was a very remarkable stroke given by Captain William Smith in this battle: with one blow of his sword he cut off the upper part of a man's skull, just under the hat. As much of the skull as was within the hat, with all the brains it contained, was struck away from the under part of it, and not so much as a fibre of the skin remained to keep them together. General Macarthy, whom James had a short time before created Lord Mountcashel, remained with five or six officers in a wood near the place of action, from which he rode out suddenly and fired a pistol on those who were guarding the artillery. A shot from one of them immediately killed his horse under him, and a musket was clubbed to knock out his brains, when he received quarter from Captain Cooper. Being asked why he hazarded his life so rashly, when he might have gone off with his cavalry, he replied that as he saw the kingdom was likely to be lost, with his own army—which, with the exception of that before Derry, then much broken, was the best in the king's service—he came upon the artillery guard with a design to lose his life, and was sorry he had missed his aim, being unwilling to outlive that day.

This was probably the greatest victory which had ever been obtained over the Irish. They amounted to 6000 men, and were routed by one-third their number. In the morning and afternoon of the day 2000 of these were killed, 500, as already mentioned, were drowned in Lough Erne, and their general, with a great many other officers, and 400 prisoners were sent to Enniskillen. The Irish confessed that 3000 of their men were wanting, when those who remained arrived in Dublin, but they would not own that so many had been killed as had been reported; in shame for having been defeated by an army so inferior in number, they alleged that the chief loss was by desertion on their retreat. They lost seven pieces of artillery, fourteen barrels of gunpowder, a great quantity of cannon and musket balls, all their drums, and every stand of colours which they possessed. The loss on the side of the Enniskilleners was only two officers, Captain Robert Corry and Ensign William Bell, with about twenty private men, who were killed. The victors would now have marched to Dublin as the Irish apprehended, to their great terror and consternation; and in all probability have carried all before them, had they not discovered, by a letter found in General Macarthy's pocket when he was taken, informing him that the Duke of Berwick, with an army from Derry, was to be at Enniskillen on a certain day, when

Colonel Sarsfield, the writer of the letter would meet it on the Connaught side with his army, then at Brendroose. The victorious army therefore returned with their prisoners and plunder to Enniskillen.

THANKSGIVING.

On the 2nd of August they went to meet Sarsfield on his way from Brendroose, but before they had got half-way, an express arrived to them from Captain Folliott, informing them that the Irish army at Brendroose had retreated to Sligo, and that the arms and ammunition intended for them by General Kirk had been landed at Ballyshannon. Three troops of horse, and as many companies of foot, were sent to besiege it, and the rest returned to Enniskillen, resolved to go in quest of the Duke of Berwick's army, in Donegal; but on the 4th of the month they heard of the relief of Londonderry, and so contented themselves with sending Lieutenant William Charleton with a troop of detached horse, to hang upon the retreating enemy's rear and watch their movements. He returned to Enniskillen in three days afterwards, and reported that he had seen the rear of them pass by Castle Caulfield, within three miles of Dungannon, on their march to Charlemont. On the 7th a solemn day of thanksgiving was observed in Enniskillen for the great victory which God had given them over their enemies, and for the peace which they enjoyed by it, after the doubts and terrors of a bloody campaign; and after divine service, the following address from the governor, officers, and clergy, and other inhabitants of the town, was drawn up and sent to King William and Queen Mary. The bearer of it was the Rev. Andrew Hamilton, Rector of Kilskeery,* in the diocese of Clogher, who, like his admirable contemporary, George Walker, was the recorder of his fellow-soldiers, as well as their counsellor in the time of doubt and suffering :—

"We, your Majesties' most faithful and loyal subjects, do in the first place offer up unto Almighty God our most humble thanks for the deliverance vouchsafed us from our merciless and

* The parish of which the Rev. W. H. Bradshaw is now rector, a clergyman who has had great acquaintance with the regiment in modern times.—[The Compiler.]

bloody enemies; and most unto your most sacred Majesties, for your gracious care taken of us, in sending Major-General Kirk * to the relief of the poor handful of your Majesties' Protestant subjects left in this place and Derry, whose miraculous holding out, under God, has been the preservation of the Protestant interest in this kingdom; and for those worthy officers sent to this place by him, among which, the Honourable Colonel Wm. Wolseley, our commander-in-chief, under whose great and happy conduct God has been pleased to bless us with the most signal and remarkable victory obtained over our enemy in this or the former age. And as we were early in the demonstration of our loyalty, in proclaiming your most sacred Majesties on the 11th of March last, so we shall persevere in the same dutiful allegiance to our lives' end, ever imploring the Divine Majesty to continue your prosperous reign long over us; most humbly begging your most sacred Majesties favourably to accept this address of our most humble and sincere obedience, which we shall ever be ready to make good both with our hearts and minds.

"GUSTAVUS HAMILTON, Governor."

[Here follow signatures.]

CLOTHING, ETC.—1747.

For the following extract from the "Regulations for the Colours, Clothing, etc., of the Marching Regiments of Foot," for 1747, I am indebted to the kindness of Lieutenant-General T. C. Smith :—

"*27th Regiment; or, The Inniskilling Regiment.*

"Allowed to wear in the centre of their colours a Castle with three Turrets, *St. George's* colours flying, in a blue Field, and the name *Inniskilling* over it.

"On the Grenadier Caps, the Castle and Name, as on the colours; White Horse and King's Motto on the Flap.

"The same Badge of the Castle and Name on the Drums and Bells of arms, Rank of the Regiment underneath.

"Facings—Pale Buff."

* An ancestor of Mrs. Freer, wife of the present colonel-commanding of the regiment.—[THE COMPILER.]

THE WATERLOO COLOURS.

An old set of the 27th's colours is suspended in Enniskillen Church. They were placed there about 1855. I find in a detached Record that on the 6th January, 1820, a new set of colours was presented to the 27th at Gibraltar. Popular belief ascribes to the church colours the merit of having been in Waterloo, and thus we have the inference that the old colours given up at Gibraltar, and which were in Waterloo in 1815, were handed over by the War Office to the town of Enniskillen. The regimental colours were subsequently changed in 1837, again in 1852, and again in 1869. Lord Enniskillen has an old set. During their stay in Enniskillen the 27th have made a happy observance. The new colours were carried to church on Sundays to do honour to the faded, moth-eaten, service-worn, dust-covered remnants of the crowning victory of Wellington's genius.

In the *Fermanagh Reporter* a series of interesting letters appeared in 1871, from the Rev. W. H. Bradshaw, now of Kilskeery, under the title of "Parochialia." He it was who presented the 27th with an address in the Main Barrack square, before their departure from Enniskillen for India. In Letter XI. he said:—

"Amongst the ornaments and donative offerings connected with our parish church I must not omit to mention the relics of the brave, which hang forth from the walls of the chancel—memorials of the men of Enniskillen and Fermanagh, who maintained the honour of their king and country at the expense of toil and blood and life. The two colours under which the 27th Inniskillings fought and bled in the numerous battles of 'the Peninsula' (the names of which are inscribed on them in full), and at Waterloo, are suspended on the northern side, under the statue of their old colonel, Sir Lowry Galbraith Cole, Bart.; and the three small flags of the 6th (Inniskilling) Dragoons, carried on the field of Waterloo, adorn the south wall of the chancel, above the monument erected in memory of the late Earl of Enniskillen. These consecrated banners, which waved amid the din of many a mortal strife, were forwarded to the Rev. Mr. Maude, as rector of the parish, by the respective commandants of these gallant corps, to hold a conspicuous place above the ashes of the honoured dead, who had valued their untarnished honour as a sacred pledge, more precious than even life itself."

THE "ROYAL" REGIMENT.

It will doubtless be surprising to many to know that the title of Royal Irish Regiment was intended for the 27th. Such at least is the testimony of Dr. O'Callaghan, at one time surgeon of the 27th, who was employed lately at the Horse Guards, where he had opportunities of obtaining information otherwise unobtainable. The following is what he says in an article in the *United Service Magazine* :—

"The title 'Royal' Regiment of Ireland was conferred upon the 18th Regiment after the siege of Namur, but it is doubtful whether this national distinction was not intended for his (William III.'s) favourite Irish regiment, the 27th Inniskilling Foot, which he always retained near his person until his death; and the grenadiers of which corps were selected by his Majesty to lead the storming party on that occasion. They were commanded by Captain Carlton, afterwards military secretary to Lord Coutts in the Peninsula, and whose memoirs present the most interesting authentic narrative of the military events of his time."

PUBLIC OPINION.

When in England lately, the military and general press spoke highly of the 27th's performances at reviews and at the autumn manœuvres. Let me make a quotation in proof of my statement:—

"The last regiment that swept past 'like a moving wall' was the old 27th, or Inniskilling Regiment, under the command of Colonel Richard Freer; their regimental colour was emblazoned with thirteen badges of distinction; their grand divisions were very strong; their lines were perfect; butts could not be seen in front of the body; every man kept his dressing, and marched straight to the front. *The French officers present did not neglect to notice* the stature and stoutness of the men."

Colonel Freer's regiment has ever won golden opinions under his command, as well as previously. The best testimony a commanding officer in his position could have as to his efficiency is the respect of those under him. Colonel Freer possesses this esteem, and has the confidence of his regiment. Colonel Freer has seen active service. When attached to the 27th in India, he "served as a volunteer with the 73rd Regiment at the action of

Budle-ke-Serai on 8th June, 1857, and in all the subsequent affairs in which the 75th Regiment was engaged during the siege of Delhi; commanded the 'Stakes' Picquet at the capture of four field-pieces at 'Ludlow Castle' on the 12th; was severely wounded near the Burn Bastion on the afternoon of the assault of Delhi (medal with clasp)."

HONOURS OF THE REGIMENT.

As marks of its meritorious service, the 27th Inniskilling Regiment bears on its appointments a Castle with the word "Inniskilling," a figure of the Sphinx with the word "Egypt" underneath, a White Horse with the words "*Nec aspera terrent,*" and the badges—St. Lucia, Maida, Nivelle, Orthes, Badajoz, Salamanca, Toulouse, Vittoria, Pyrenees, Peninsula, and Waterloo. The Castle is always depicted with three turrets, and St. George's colours flying. It was taken from the regiment by certain influence in the latter end of the seventeenth or in the eighteenth century, but it was restored by one of the Georges. The same monarch would have given the title of "Royal" Regiment to the 27th, on the grounds stated by Dr. O'Callaghan, but that he considered its deprivation from a regiment which had borne it so long would not be agreeable to either of the regiments. The Sphinx tells its own tale. The White Horse speaks of ancient service like the Castle. The White Horse is a Hanoverian badge. In the officers' mess is preserved a copy of a letter written in French by King William III. during the siege, and not in the best of French either, I may say. I give a translation below. The siege of Namur took place in the year 1695, but the error may be the fault of the copyist. The 27th is believed to be one of the regiments the king had placed between the rivers Sambre and Meuse. The letter speaks of the opening of the trench on the night of the 12th; Wm. Howitt places it on the 11th, and Macaulay on the 2nd. Whether this letter be a copy of the one intended to delude the French or not, or to one of his generals, I cannot say.

"*At the camp before Namur, this* 12*th July*, 1693, *at six o'clock in the evening.*

"I received on this morning your two letters of yesterday ten o'clock in the evening. I am rather embarrassed as to how I am to send you the three regiments of dragoons which you would like to have, as they are all scattered in different positions and very useful where they are. The Spanish, which you know are very feeble, and the two Bavarian regiments are between the Sambre and the Meuse, and this is the only cavalry which the Elector has at his command. It is therefore not likely that he will be able to deprive himself of it. I have between the Meuse and the Sambre two regiments which are for all the convoys and parties. I have another one which is placed between Huy and the army for the escort of the battalions; and the Count of Athlone has four regiments of dragoons with him, which he is most in need of to cover his camp against so much infantry which are rambling about him. Notwithstanding all this, I shall reduce it to three, if you think yourself absolutely in need. If the march was not so long I should have sent you three battalions of infantry, and if you wish I shall send them to you, but I expect first an answer from you. In case of extreme necessity you might take the garrisons of Ath and Dykk, and the battalion which I am able to send from here could replace them. We have opened this night the trench, which took place very peaceably, but the reason of this may be that it had been opened in the distance. Dopp Masseine will inform you daily of everything that may happen, so do but things which are worth the trouble.

"I am yours always,
(Signed) "W. R."

PENINSULAR OFFICERS.

Names of officers of different battalions who, in consideration of their services and conduct in engagements with the enemy in Spain, Portugal, and France, were distinguished with titles, medals, and with other marks of his Majesty's favour:—

Colonel Maclean, commanding the 3rd Battalion in all the campaigns in the Peninsula and France; Knight Commander of the Bath; a cross and two clasps for Salamanca, Vittoria, Pyrenees, Nivelle, Orthes, and Toulouse. The order of the Tower and Sword was also conferred upon him by the King of Portugal.

Lieutenant-Colonel George James Reeves, commanding 2nd Battalion in Spain; Companion of the Bath; Knight of the Guelphic Order.

Lieutenant-Colonel John Hare was promoted to the brevet

rank of major for his services with the 2nd Battalion in Spain, and to that of lieutenant-colonel, for commanding the 1st Battalion at Waterloo; Companion of the Bath; a Waterloo medal. The order of St. Vladimir was also conferred upon him by the Emperor of Russia.

Lieutenant-Colonel Henry Thomas, afterwards of the 20th Regiment, was promoted to the brevet rank of major, and afterwards to that of lieutenant-colonel, for his services and conduct in the Peninsula in command of the light companies of the brigade in the 4th Division, with which the 3rd Battalion 27th Regiment acted; Companion of the Bath; a medal and two clasps for Nivelle, Orthes, and Toulouse.

Brevet-Lieutenant-Colonel J. R. Ward, major half-pay 36th Regiment; Companion of the Bath; medal and two clasps for Salamanca, Badajoz, and Pyrenees.

GENERAL COLE.

On the fort erected by the Inniskilliners in 1688-9, now known as the Fort hill, convenient to the town of Enniskillen, is erected a monument to General Cole, over 100 feet in height. A statue of the general is on the summit. A winding staircase of 109 steps leads to the platform.

On three sides of the base are engraved the following words:—

Martinique	Olivença	Pyrenees
Guadaloupe	Albuera	Nivelle
Egypt	Salamanca	Orthes
Maida	Vittoria	Toulouse

Over the entrance door is the following inscription:—

IN MEMORY OF

GENERAL THE HON[BLE] SIR GALBRAITH LOWRY COLE, G.C.B.,

COLONEL OF THE 27TH REGIMENT.

ERECTED BY HIS FRIENDS,

1843.

On an ornamented copper plate inside the structure is the inscription below:—

THIS PILLAR IS ERECTED BY HIS FRIENDS AND
FELLOW-COUNTRYMEN IN MEMORY OF

GENERAL THE HON^BLE SIR G. LOWRY COLE,

KNIGHT GRAND CROSS OF THE ORDER OF THE BATH;
KNIGHT OF THE PORTUGUESE ORDER OF THE TOWER AND SWORD,
AND OF THE TURKISH ORDER OF THE CRESCENT;
COLONEL OF THE 27TH INNISKILLING REGIMENT OF FOOT;
GENERAL COMMANDING THE
4TH DIVISION OF THE BRITISH ARMY
THROUGHOUT THE PENINSULAR WAR;
GOVERNOR OF GRAVESEND AND TILBURY FORT;
M.P. IN THE IRISH HOUSE OF COMMONS FOR THE
BOROUGH OF ENNISKILLEN FROM 1798 TO 1800,
AND IN THE IMPERIAL PARLIAMENT FOR THE
COUNTY OF FERMANAGH FROM 1803 TO 1823.
HE TWICE RECEIVED THE THANKS OF BOTH HOUSES
OF PARLIAMENT FOR HIS DISTINGUISHED MILITARY SERVICES.
BORN MAY 1ST, 1772; DIED OCT. 4TH, 1842.

THE STATUE IS THE CONTRIBUTION OF THE TENANTRY
OF THE ENNISKILLEN ESTATES.

COLONEL REEVES AND CASTALLA.

The following copy of a letter I procured at the Horse Guards will explain itself:—

"Before Barcelona, 8th April, 1814.

"SIR,

I consider myself particularly happy in being deputed by the officers of the 2nd Battalion 27th Regiment to request your acceptance of a sword, value one hundred guineas, as the most honourable mark we can offer of our admiration of your gallant conduct at our head, in the actions of Biar and Castalla on the 12th and 13th April, and the Col-de-Ordal on the 13th September, 1813; as well as in grateful remembrance of the uniform happiness we have at all times experienced in performing our duties under your command. And I beg to add, that I feel proud in being made the channel of conveying to you this well-merited proof of the high sense we entertain of the increased reputation reflected on the corps by the conduct of its brave leader on the above occasions. That you may long live to wear this badge of our respect and esteem is the warmest wish of,

"Yours most faithfully,
(Signed) "JOHN HARE,
"Major commanding 2nd Battalion 27th,
or Enniskillens.

"Lieutenant-Colonel Reeves,
"2nd Battalion Enniskillens."

27TH INNISKILLINGS.

NOMINAL ROLL OF THE OFFICERS OF THE ABOVE CORPS.

October, 1875.

RANK.	NAME.
Colonel	R. Freer.
Major	H. Cowell.
"	A. D. Geddes.
Captain	W. Herring.
"	H. B. P. Phillipp.
"	W. Pott.
"	A. Hales.
"	W. R. Tredennick.
"	D. M. Taylor.
"	H. M. Caine.
"	J. M. Kerr.
"	F. Coffey.
"	R. W. E. White.
"	R. W. Brownrigg.
Lieutenant	J. W. F. Buxton.
" and Ins. Musk. ...	P. Stainforth.
"	E. Barton.
"	W. W. Kettlewell.
" and Adjutant ...	H. Wodehouse.
"	C. W. Hare.
"	A. J. Irvine.
"	G. H. Michaelson.
"	A. H. Grant.
"	J. C. Bayly.

RANK.	NAME.
Sub-Lieutenant	A. G. M. Bond.
"	H. S. Tunnard.
"	J. A. Bennett.
"	H. P. Williams.
"	A. H. Young.
"	N. A. Bray.
Quartermaster	C. Brennan.
Surgeon-Major	A. Morphew.

SUCCESSION OF COLONELS, 27TH REGIMENT,

With Date of Appointment.

Zachariah Tiffin	26th June, 1689.
Thomas Whetham	29th August, 1702.
Richard Molesworth	22nd March, 1725.
Archibald Hamilton	29th May, 1731.
Sir W. L. Blakeney, K.B.	22nd June, 1737.
Hugh Warburton	26th September, 1761.
Sir Eyre Coote, K.B.	5th September, 1771.
Eyre, Lord Clarina	19th February, 1773.
F. Marquis of Hastings, K.G., G.C.B., G.C.H.	23rd May, 1804.
Hon. Sir Galbraith Lowry Cole, G.C.B.	16th December, 1826.
Sir John Maclean, K.C.B.	2nd November, 1842.
Sir W. Francis Patrick Napier, K.C.B.	5th February, 1848.
Edward Fleming, C.B.	18th September, 1853.
John Geddes, K.H.	24th April, 1860.
James Robertson Craufurd ...	29th April, 1864.
Randal Rumley	27th August, 1870.

MEMOIR OF THE SERVICES OF MAJOR-GENERAL L. WARREN.*

In 1787 this officer entered the army as an ensign in the 17th Foot, in which corps he obtained a lieutenancy in 1789 and in the latter year embarked with his regiment on board Admiral Lord Hood's fleet, where they were ordered to serve as marines.

In 1793 he raised an independent company, and in the following year exchanged into the 27th Regiment, then forming part of Lord Moira's army, encamped at Southampton. The critical situation of the Duke of York in Flanders at this period occasioned his lordship to be despatched with a reinforcement of 10,000 men to aid his Royal Highness, with whom, though nearly surrounded by much superior armies in point of numbers, Lord Moira, by a well-directed movement, effected a junction near Malines, and thus relieved the British army from the difficulties of its situation, to the mortification of the French general, Pichegru. In this well-conducted expedition Captain Warren served with the 27th. He was also present at the siege of Nimeguen, the sortie on the evening of the 6th of November, and commanded the advanced picquet of the garrison. In December he accompanied the forces under Lord Cathcart, sent to attack the French army that had crossed at Bonomell, and was present in the action of Geldermalsen, the 8th of January, 1796.

The 27th Regiment embarked in September, 1796, for the West Indies, and Captain Warren was accordingly present at the siege of Morne Fortuné, St. Lucia, and commanded the grenadiers at the storming of the enemy's advanced posts; at the conclusion of which service he was compelled by sickness to return, on leave, to England.

In 1799 he served in the expedition to the Helder, and was engaged in the actions of the 27th of August, 19th of September, 2nd and 6th of October.

In August, 1800, this officer, then senior major of the 1st Battalion 27th Foot, served in the expedition to Ferrol. In September following, the 1st Battalion joined Sir Ralph Abercromby's expedition before Cadiz; it afterwards proceeded to Malta, where it was disembarked in consequence of sickness.

* Major-General Warren was great-uncle of Mr. Barton, 27th Inniskillings.

In April, 1801, Major Warren sailed with the battalion for Egypt, and was employed with it on the whole service against Alexandria, from the beginning of May till the surrender of the place; the battalion forming, on the 27th of August, General Sir Eyre Coote's advanced guard, on his approach to Alexandria on the western side.

In 1804 this officer became lieutenant-colonel in the 27th Regiment; and in February, 1806, he embarked with it for Hanover, from whence he returned in April following. He next embarked for Sicily, and was in the expedition to the Bay of Naples, under General Sir John Stuart. From August, 1809, when Sir John Stuart returned from the Bay of Naples, until November, 1812, Lieutenant-Colonel Warren continued in Sicily. He afterwards embarked with the 1st Battalion of his regiment for the eastern coast of Spain, where he was immediately appointed to the command of a brigade, with which he served at the battle of Castalla, the 13th of April, and at the siege of Tarragona. In the following year he was at the blockade of Barcelona. Colonel Warren accompanied the division of the British army across the Peninsula to Bayonne, and from thence to Bordeaux, where the 27th was immediately embarked for North America. He then obtained leave of absence, but in the following year, 1815, joined the 1st Battalion of the 27th Regiment before Paris, a few days prior to the entrance of Louis XVIII.

In 1819 this officer obtained the brevet of major-general. He maintained throughout his career the character of a brave and skilful regimental officer. The major-general died suddenly in London, on the 29th of October last.—From the *United Service Journal*, January, 1834.

THE FOLLOWING TABLE SHOWS THE SERVICES OF SOME OF THE LATE AND PRESENT OFFICERS OF THE REGIMENT.

CORPS.	DATES.	RANK AND NAMES.	BATTLES, ETC.	WOUNDS.	MEDALS.	REMARKS.
R. H. Gds. 27th Regt.	1 Oct., 1812 18 April, '43	Lieut.-Col. H. R. Magennis	Served with 7th Fusiliers in Spain and France; was present at Nivelle, Orthes, and Toulouse, under the command of his Grace the Duke of Wellington; served with the same regiment in America, and present at the battle of New Orleans under the command of E. Pakenham.	...	Silver medal and three clasps for Nivelle, Orthes, and Toulouse.	Lieutenant. Lieut.-Colonel.
†60th Rifles 27th Regt.	2 Nov., 1830 2 April, 1852	†Lieut.-Col. A. A. Cunynghame	Captain 3rd Buffs; Aide-de-camp to Maj.-Gen. Lord Soulton at the siege of Chan-Kiang-Foo, and the investment of Nankin, in the year 1842.	...	Medal for service in China.	* 2nd Lieutenant. Served in Gibraltar, Corfu, China, and North America, from 1835 to 1851.
‡97th Regt. 27th Regt.	21 June, '38 12 Nov., '52	Capt. J. R. G. Patteson	Campaign of 1845-46; Army of the Sutlej; present at battle of Sobraon, 10th Feb., 1846, and occupation of Lahore as Lt., His Excellency Gen. Sir Hugh Gough, G.C.B., Com.-in-chief in India, commanding the army in person; served in the campaign of 1848-49, including the whole of the siege operations against Mooltan, surrender of the fortress, 22nd January, 1849, battle of Goojerat, 21st Feb., 1849, as Lieut., under command of Gen. Lord Gough.	...	Silver medal of Sobraon, Punjab medal, and two clasps inscribed Mooltan, Goojerat, for the campaign of 1848-49.	‡ Captain.
12th Foot 27th Foot	12 Jan., '49 10 July, '60	Major H. Cowell	Advance on Delhi and Kurnal; battle of Budle-ke-Serai; operations before Delhi.	...	Indian Mutiny medal, and clasp for Delhi.	Promoted for service at Delhi to Capt. in 2-14th Regt.
28th Regt. §27th Foot	27 Sept. '33 22 July, '59	Colonel R. J. Baumgartner	Served in Eastern campaign of 1854-55, battles of Alma, Inkermann, siege of Sebastopol; succeeded to the command of the 28th Regt. in the attack and oc-	...	Crimean medal, with clasps for Alma, Inkermann, and Sebastopol; C.B.;	§ Lieut.-Colonel.

			cupation of the Cemetery on 18th June, 1855, and brought the 28th Regt. out of action.		fourth class of the Medjidié; Sardinian and Turkish medals.	
Hospital Assistant 27th Regt.	8 Nov., 1810 19 Dec., '11	Surgeon Thomas Mostyn	Siege of Badajoz, April, 1812; affair of Caraval, 18th July, 1812; Salamanca, 22nd July, 1812; Vittoria, 21st March, 1813; action in Pyrenees, August, 1813; assault of San Sebastian, 31st August, 1813; Nivelle, 10th Nov., 1813; Nive, Dec., 1813; Orthes, Feb., 1814; sortie from Bayonne, March, 1814; advance to Plattsburg, 1814; Waterloo, June, 1815; capture of Paris, 1815.	...	Waterloo medal, and silver war medal and eight clasps for Badajoz, Salamanca, Vittoria, Pyrenees, San Sebastian, Nivelle, Nive, Toulouse.	Noticed in G. O. dated Gibraltar, March, 1823, for service rendered to the Emperor of Morocco.
27th Regt.	1 Dec., 1837	Brevet-Major E. N. Molesworth	Port Natal, as Adjt., in 1842; Kaffir war, 1846-47, as Field-Adjt. to 1st Division, under Col. Hare, Lieut.-Gov. Eastern Frontier during Mutiny of 1857-58.	...	Kaffir war medal, South Africa.	Promoted to Brevet-Major for conduct on board the ship *Eastern Monarch*.
60th Rifles 27th Regt.	22 May, 1846 4 April, 1856 19 Jan., 1858 18 Oct., 1859 27 Dec., 1868 25 June, '69 23 June, '74	Col. Richard Freer Captain Bt.-Major Major Bt.-Lieut.-Col. Lieut.-Col. Colonel	Indian Mutiny, 1857; Budle-ke-Serai sorties; assault on Delhi.	Wounded nr. the Burn bastion on afternoon of assault on Delhi, 14th Sept., 1857. Received £100 for wound.	Indian Mutiny medal, and clasp for Delhi.	Mentioned in despatches of Brigadier Showers and Maj.-Gen. Sir A. Wilson; filled the offices of Deputy Assistant Quartermaster-General of the Army, head-quarters Calcutta, and Deputy Assistant Quartermaster-General, Presidency Division Bengal Army, which he vacated on promotion to Regimental Major.

"THE INNISKILLINGS."

A SONG OF THE GALLANT 27TH.

Inscribed to Mrs. Freer, and composed by W. Copeland Trimble.

Being an endeavour to epitomize in verse a record of the most prominent deeds of the regiment.

I.

Come, listen, my boys, and I'll give you a song,
Come, join in the chorus and sing it out strong,
Of our own Twenty-seventh, of fame and renown,
The red-coated boys of our darling old town.
Their banners look'd down on Culloden's red plain—
They oft saw the Frenchies in Canada slain—
St. Lucia her tricolor haul'd to the clay
For the brave Inniskillings' buff banner's display.

Chorus.

Then hurrah for the victors of many a field,
Hurrah for the heroes that know not to yield,
For the brave Inniskillings a welcome we'll make,
When they come from afar to their home by the lake.

II.

When Holland was still the great Emperor's aim,
Sir Ralph Abercromby defeated his game,
For the brave Inniskillings put "Frenchy" to flight,
When he vainly essay'd to imperil our "Right."
Napoleon's ambition on Afric next lay,
Our boys went to meet him at Aboukir Bay,
And "Egypt" and "Maida" to this day proclaim
How well they remembered their ancestors' name.

Chorus.—Then hurrah, &c.

III.

When Bony his brother would make King of Spain,
Our fellows soon thrash'd him in battle again.
Castalla's green hillsides and Suchet could tell
How a Waldron could fight for the cause he loved well.
Spain's rills and sierras received their hearts' blood,
Salamanca's dark woods can proclaim how they stood,
San Sebastian's walls heard their death-pealing cry,
Toulouse's proud heights saw their banners on high.

Chorus.—Then hurrah, &c.

IV.

When Waterloo quail'd 'neath the Emperor's host,
Before whom the armies of Europe were toss'd,
He tried, but in vain, to dislodge gallant Hare,
Though four hundred fell in his valorous square.
We hear of the deeds at the Cape they have done,
And of hardships untold beneath India's sun;
They toil'd and endured when the Sepoy rebell'd,
Till Delhi's bold seizure the Mutiny quell'd.

Chorus.—Then hurrah, &c.

V.

Once again are they home on their own native sod,
Yet from Inis-Kethlenn * must they wander abroad,†
But before they depart will we cheerily cheer
For our own Twenty-seventh and brave Colonel Freer!
And hurrah for the majors and officers all,
For Windrum and sergeants who come at his call!
Hurrah for Herr Werner, for Presley, and band,
And hurrah, boys, hurrah for our dear native land!

Chorus.—Then hurrah, &c.

* Pronounced Inis-kellen.

† Referring to their approaching departure to India.

*** For references of the song see the Record. Words and music can be had from the Author, 7, East Bridge Street, Enniskillen, post-free for 25 stamps.

www.ingramcontent.com/pod-product-compliance
Ingram Content Group UK Ltd.
Pitfield, Milton Keynes, MK11 3LW, UK
UKHW021837270726
14058UKWH00002B/195